Springer Theses

Recognizing Outstanding Ph.D. Research

Aims and Scope

The series "Springer Theses" brings together a selection of the very best Ph.D. theses from around the world and across the physical sciences. Nominated and endorsed by two recognized specialists, each published volume has been selected for its scientific excellence and the high impact of its contents for the pertinent field of research. For greater accessibility to non-specialists, the published versions include an extended introduction, as well as a foreword by the student's supervisor explaining the special relevance of the work for the field. As a whole, the series will provide a valuable resource both for newcomers to the research fields described, and for other scientists seeking detailed background information on special questions. Finally, it provides an accredited documentation of the valuable contributions made by today's younger generation of scientists.

Theses may be nominated for publication in this series by heads of department at internationally leading universities or institutes and should fulfill all of the following criteria

- They must be written in good English.
- The topic should fall within the confines of Chemistry, Physics, Earth Sciences, Engineering and related interdisciplinary fields such as Materials, Nanoscience, Chemical Engineering, Complex Systems and Biophysics.
- The work reported in the thesis must represent a significant scientific advance.
- If the thesis includes previously published material, permission to reproduce this must be gained from the respective copyright holder (a maximum 30% of the thesis should be a verbatim reproduction from the author's previous publications).
- They must have been examined and passed during the 12 months prior to nomination.
- Each thesis should include a foreword by the supervisor outlining the significance of its content.
- The theses should have a clearly defined structure including an introduction accessible to new PhD students and scientists not expert in the relevant field.

Indexed by zbMATH.

Jiangfeng Lu

Novel Proton Conducting in Free-Standing Coordination Polymer Membranes

Doctoral Thesis accepted by
Kyoto University, Kyoto, Japan

Springer

Author
Dr. Jiangfeng Lu
Fujian Institute of Research on
the Structure of Matter
Fuzhou, China

Supervisor
Prof. Hiroshi Kitagawa
Division of Chemistry
Graduate School of Science
Kyoto University
Kyoto, Japan

ISSN 2190-5053 ISSN 2190-5061 (electronic)
Springer Theses
ISBN 978-981-97-7912-3 ISBN 978-981-97-7913-0 (eBook)
https://doi.org/10.1007/978-981-97-7913-0

This Springer imprint is published by the registered company Springer Nature Singapore Pte Ltd.
The registered company address is: 152 Beach Road, #21-01/04 Gateway East, Singapore 189721, Singapore

Supervisor's Foreword

The doctoral thesis of Dr. Jiangfeng Lu was completed at Kyoto University's Graduate School of Science and was nominated as a Springer Thesis due to its exceptional contribution to the field and high level of originality. In 2020, Dr. Lu joined my research group as a Ph.D. student. Our team has been actively studying proton conduction in porous coordination polymers (PCPs) to create high-performance proton-conducting materials. Dr. Lu has dedicated his efforts to studying proton conduction in coordination polymers, uncovering several fascinating systems and phenomena. He has summarized these findings into a paper that highlights the unique behavior of proton conduction in PCP membranes.

This thesis addresses three main themes: First, the creation of heterojunctions from hydroxide and proton ion conductors to facilitate unidirectional proton conducting; second, a meticulous exploration of the proton conductivity in self-supporting membranes, emphasizing their mechanical flexibility against bending stress and robust proton conduction against mechanical stress; and third, the synthesis of flexible coordination polymer free-standing membranes and an analysis of the variations in their proton conduction properties. Dr. Lu's innovative work has been recognized in the esteemed publication, *Angewandte Chemie International Edition*. His contributions have been acknowledged by peers, with one of his papers being hailed as a *Very Important Paper* and *Cover Paper*. This recognition underscores the thesis's nomination for the Springer publication, highlighting its exceptional scholarly merit.

Kyoto, Japan
July 2024

Prof. Hiroshi Kitagawa

Parts of this thesis have been published in the following journal articles:

1. **High-Performance All-Solid-State Proton Rectifier Using a Heterogeneous Membrane Composed of Coordination Polymer and Layered Double Hydroxide**

   Jiangfeng Lu; Yukihiro Yoshida; Mitsuhiko Maesato; Hiroshi Kitagawa, *Angew. Chem. Int. Ed.* **2022**, *61*, e202213077.
2. **Robust Proton Conduction against Mechanical Stress in Flexible Free-Standing Membrane Composed of Two-Dimensional Coordination Polymer**

   Jiangfeng Lu; Yukihiro Yoshida; Kazuyoshi Kanamori; Hiroshi Kitagawa, *Angew. Chem. Int. Ed.* **2023**, *135*, e202306942.

Acknowledgements

The presented thesis summarizes the author's study from October 2020 to September 2023 under the supervisor, Prof. Hiroshi Kitagawa (Division of Chemistry, Graduate School of Science, Kyoto University).

I extend heartfelt gratitude to my esteemed primary supervisor, Prof. Hiroshi Kitagawa, for the invaluable opportunity to pursue my Ph.D. within the current laboratory. His unwavering support throughout my doctoral journey has been a constant source of strength and an inspiration. Professor Kitagawa's paternal figure has been a beacon of encouragement and support in times of adversity. It is with great honor that I have been able to learn from his exemplary character and tireless work ethic, qualities that I shall cherish and carry with me throughout my life. The successful completion of this thesis is a testament to his outstanding mentorship, the profound insights he offered into my research, and the continuous encouragement that propelled me forward.

Secondly, I want to express my deep gratitude to my direct supervisor Dr. Yukihiro Yoshida. He is the most responsible teacher I have ever met. His work attitude has also deeply influenced me and will benefit me for the rest of my life. He provided me with comprehensive and detailed guidance for completing the paper. His consistent excellent guidance and support helped me overcome the problems and difficulties encountered in the project research, saving me many detours in my daily work and study. This thesis could be carried out well because of his sincere and selfless support, prompt and valuable suggestions, and kind encouragement.

I have been fortunate to receive exceptional support in both experimental work and intellectual discourse from Dr. Mitsuhiko Maesato and Dr. Kazuyoshi Kanamori at Kyoto University. It is with sincere gratitude that I acknowledge their invaluable contributions. The thoughtful suggestions and enriching discussions they have provided have been instrumental in my research process.

I extend my deepest gratitude to Dr. Quan Zhang and Dr. Dongshuang Wu. Their assistance has been invaluable, both in my professional endeavors and personal life. Their support has been a guiding light, helping me navigate through various challenges and enriching my experiences.

I would like to extend my special thanks to the entire team at Prof. Hiroshi Kitagawa's laboratory. Their collective wisdom, encouragement, and camaraderie have been invaluable assets throughout my research journey. Each member's contribution, through insightful suggestions and unwavering support, has enriched my experience and enhanced the quality of my work.

I offer my heartfelt thanks with the utmost sincerity, profoundly aware that my accomplishments would not have been possible without the unwavering love and nurturing care of my family and the many others who have played a part in my upbringing. Their collective influence has been the bedrock of my successes and a source of constant inspiration.

Contents

Chapter 1
General Introduction

Abstract This section delves into the groundbreaking discovery of proton conduction in free-standing coordination polymer membranes (PCPs). These PCPs are crafted through the intricate self-assembly process of metal ions with organic linkers, resulting in exceptional crystallinity and adjustable porosity. The study zeroes in on the pivotal Grotthuss and Vehicle mechanisms for proton transfer, meticulously evaluating their conductivity using AC impedance spectroscopy and single-crystal analysis. The innovation is underscored by the development of supple yet robust PCP membranes, designed to uphold their performance even under mechanical strain, thereby highlighting their immense potential for application in the next wave of protonic devices.

Keywords Coordination polymer membranes · Proton conduction · Grotthuss mechanism · Proton rectification · Metal–organic frameworks (MOFs)

1.1 Proton-Conducting Porous Coordination Polymers

Porous coordination polymers (PCPs) or metal–organic frameworks (MOFs) are porous materials with periodic network structure. It is formed by self-assembling metal-containing (metal ions or metal clusters) and organic linkers (Fig. 1.1), which shows high crystallinity, designability, porosity, and rich internal channels [1]. By regulating the types and coordination modes of metal and organic ligands, PCPs can exhibit specific structures and properties, which make them widely used in essential fields such as catalysis [2], sensing [3], gas storage, and separation [4]. Because the judicious selection of components can control the surface potential (i.e., hydrophilicity/hydrophobicity and Lewis acidity/basicity) and geometrical structure of the pore spaces, PCPs provide a versatile platform for proton conduction when Lewis acidic molecules (e.g., phosphoric acid) or Lewis basic molecules (e.g., water) are accommodated in the pores [5]. Therefore, PCPs have great potential for protonic applications, such as proton-exchange membrane fuel cells [6], proton field-effect transistors [7], hydrogen sensors [8], and proton separation [9].

J. Lu, *Novel Proton Conducting in Free-Standing Coordination Polymer Membranes*, Springer Theses, https://doi.org/10.1007/978-981-97-7913-0_1

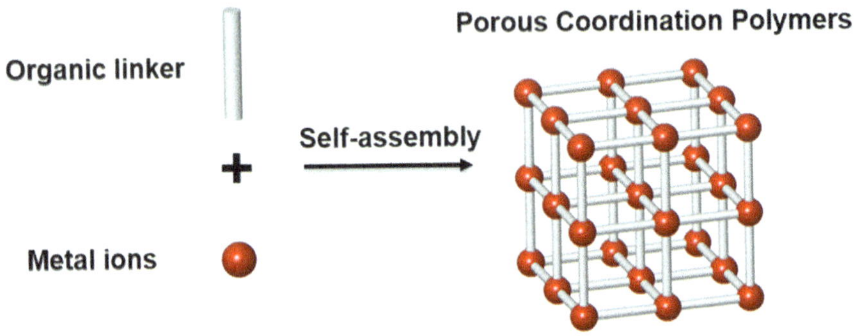

Fig. 1.1 Synthesis schematic of porous coordination polymers

Studies on PCPs' proton conductors began in 1979, when Kanda et al. [10] firstly confirmed the proton conductivity of $(HOC_2H_4)_2$-dtoa-H_2 two-dimensional structure by the lamellar powder method. Since then, people have begun to recognize the potential of PCPs as proton conductors and continue to design and develop new PCPs' structures with excellent proton transfer ability. The advantages of PCPs applied in the proton-conducting field are as follows: [11] (1) The porous property of PCPs enables them to effectively adsorb proton carriers and provide effective active sites for proton transfer; (2) the regularity of PCPs' pore structure leads to the formation of efficient proton transfer paths; and (3) the straightforward synthesis of PCPs makes it possible to prepare proton conductors that meet specific needs on a large scale.

It is widely believed that there are two main mechanisms of proton transport in PCPs: Grotthuss mechanism and Vehicle mechanism (Fig. 1.2) [12]. The process of proton jumping mechanism is as follows: First, the adjacent molecules or groups inside the proton conductor form a large-scale hydrogen bond network; the protons then combine with water molecules to form hydronium ions; finally, driven by the chemical potential, the proton jumps to the next adjacent molecule or group through the hydrogen bond network, completing the proton migration and transfer. The proton transport mechanism relies on the diffusion of the proton carrier inside the conductor to complete the transmission process. In this process, the proton-carrying carriers move along the transmission direction, and the concentration gradient will be formed in the conductor after the diffusion of these carriers, making the remaining carriers without proton-carrying move in the opposite direction. The proton jumping mechanism only involves proton transfer, and the required activation energy (E_a) is generally less than 0.4 eV [14]. However, the proton transport mechanism involves the diffusion of a larger proton carrier and requires a larger activation energy ($E_a > 0.4$ eV) [15]. Grotthuss ($E_a < 0.4$ eV) and Vehicle mechanisms ($E_a > 0.4$ eV) have a distinct E a range, which is evaluated.

$$\sigma T = A \exp\left(-\frac{E}{k_B T}\right)$$

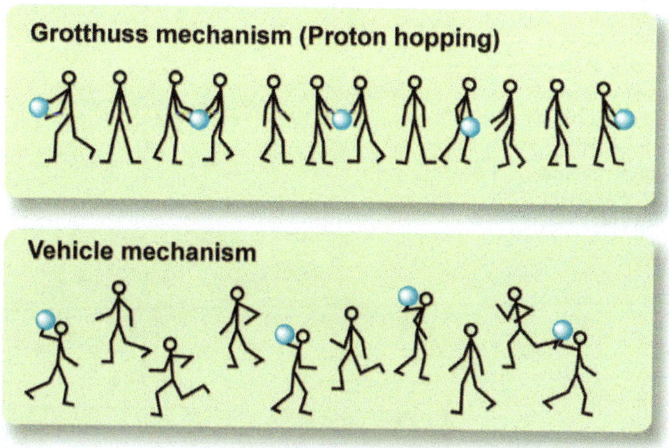

Fig. 1.2 Schematic illustration of Grotthuss and Vehicle mechanisms in proton conducting. Reprinted with permission from Ref. [13]. Copyright 2008 American Chemical Society

by fitting the temperature-dependent conductivity to the Arrhenius equation.

Where σ is the ion conductivity, T is the temperature, A is the Arrhenius factor, and kB is the Boltzmann constant. The linear fit of $\ln(\sigma T)$ versus $T - 1$ yielded E from the corresponding slope.

It is worth mentioning that due to the complex operating environment of the actual proton conductor, proton transmission often relies on the combination of the two mechanisms.

AC impedance method is commonly used to study the proton conductivity of PCPs. In this method, the synthesized PCPs' powder is generally pressed into a disk of a certain size and thickness for electrochemical testing, and the action mechanism of PCPs proton conductor is analyzed in combination with other characterization methods [16]. However, the conductivity of PCPs measured by this method is greatly affected by grain boundaries. In order to study the nature of proton conduction of PCPs in detail, more people use single-crystal measurement method. The biggest advantage of single-crystal method is that it can measure the proton transport performance in all directions of PCPs crystal, which makes people realize that the proton conduction of PCPs has anisotropy. Kitagawa et al. [17] prepared a two-dimensional structure coordination framework compound $Zn(H_2PO_4)_2(TzH)_2$ (Fig. 1.3a). The anhydroion conductivity of this PCPs is 1.51×10^{-3} S cm^{-1} (150 °C, $E_a = 0.6$ eV), and the single-crystal test results show (Fig. 1.3b) that the conductivity of this crystal parallel to *ab* plane is much higher than that perpendicular to *ab* plane *c* direction. It is proved that hydrogen bond network in *ab* plane is the source of proton conductivity of PCPs. In addition, characterization methods such as quasi-elastic neutron scattering [18], fluorescence scattering [19], solid-state NMR [20], as well as analytical and computational techniques such as molecular dynamics simulation [21] and

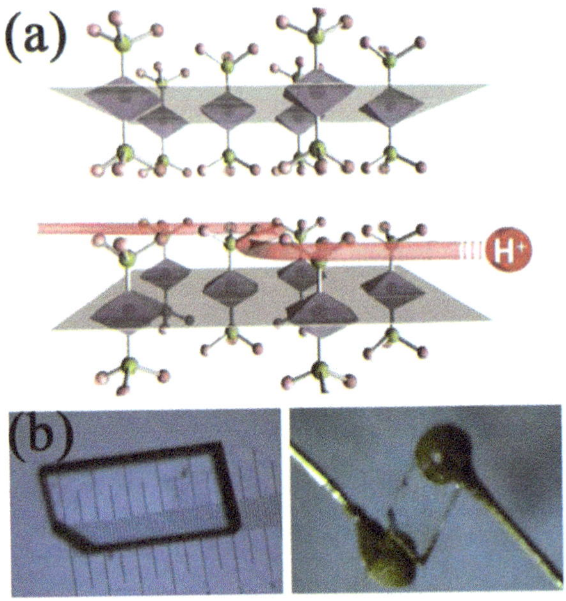

Fig. 1.3 **a** Structure schematic illustration of proton conducting, **b** single crystal (left) and with gold electrolytes attached (right) of $Zn(H_2PO_4)_2\ (TzH)_2$. Reprinted with permission from Ref. [17]. Copyright 2012 American Chemical Society

density functional theory [22] make it possible to directly understand the motion of protons in PCPs.

1.2 Porous Coordination Polymer Membranes

PCP materials have unique physical and chemical properties and have a wide range of application prospects. Applying the porous, highly ordered of PCP in thin membrane field will create them with better performance, for example (Fig. 1.4 [23]), catalysis [24], adsorption [25], separation [26], electricity [27], optics [28], and sensor [29]. At present, the research of PCP membranes mainly focuses on the synthesis, preparation, and properties of membrane with substrate support and self-supporting membranes. However, how to arrange the pores of PCP in a highly ordered manner in the process of membrane formation without the help of substrate is still an unsolved problem in this field for more than 20 years. At present, the research on PCP membranes is mainly divided into three categories, namely, the synthesis, preparation, and performance research of membranes with substrate support, PCP-based membranes, and self-supporting membranes.

The first type is the membranes with substrate support. The membranes with substrate support are a kind of membrane grown or attached to the substrate preparation; the selection of the substrate is various, including flexible substrate and rigid substrate [30]. By modifying the substrate, the binding force between the substrate and the membrane can be improved, resulting in the membrane being closely

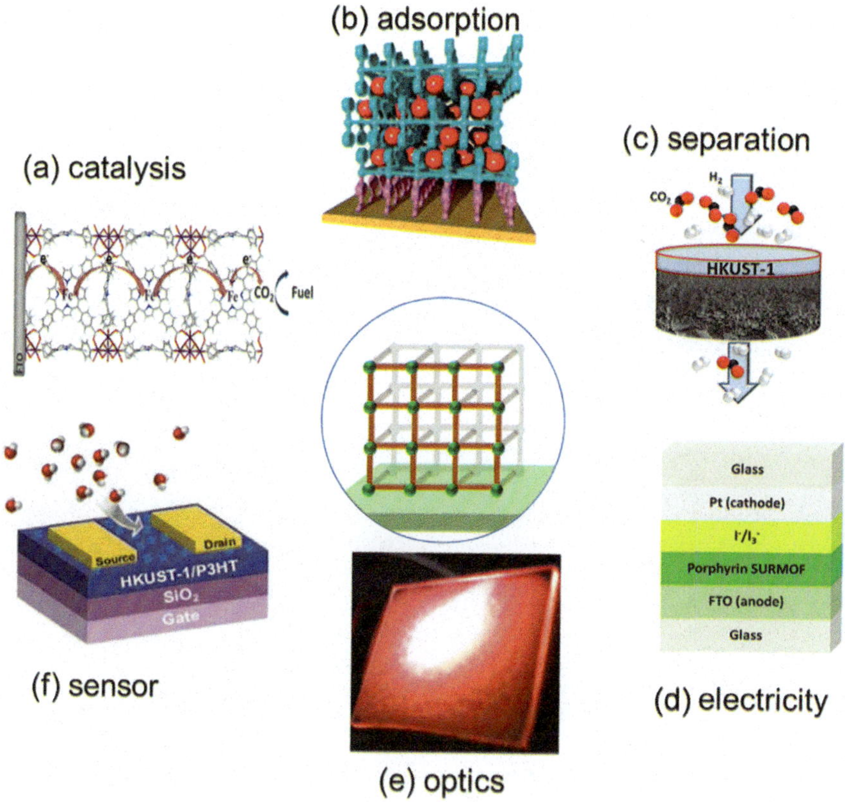

Fig. 1.4 Applications of PCP membranes. Reprinted with permission from Ref. [23]. Copyright 2021 Elsevier

attached. Currently, the main synthesis methods reported include in situ synthesis method [31], LB method [32], spin-coating method [33], layer self-assembly method [34], seed crystal method [35], etc. However, they also have shortcomings such as the limitation of the use of membranes by the substrate and certain requirements on the substrate. As shown in Fig. 1.5, Xu et al. selected flake PCP crystals and conceived a novel film-making method, stamp method [36]. The method is to make CuTCPP nanosheet (diameter 300–500 nm) dispersed in ethanol to form colloidal solution, take a small amount of colloidal solution dispersed on the surface of water. Hydrophobic CuTCPP will spontaneously form a film on the surface of water, and then use the substrate to dip the film from the liquid surface, thus forming crystalline CuTCPP film on the substrate. The thickness of the film can be controlled by the number of dip-cycles, and the relationship between the thickness of the film and the number of cycles can be characterized by the ultraviolet absorption intensity.

The second type is to embed PCPs in a polymer matrix substrate. The thickness of polymer is much larger than the diameter of the embedded porous PCPs.

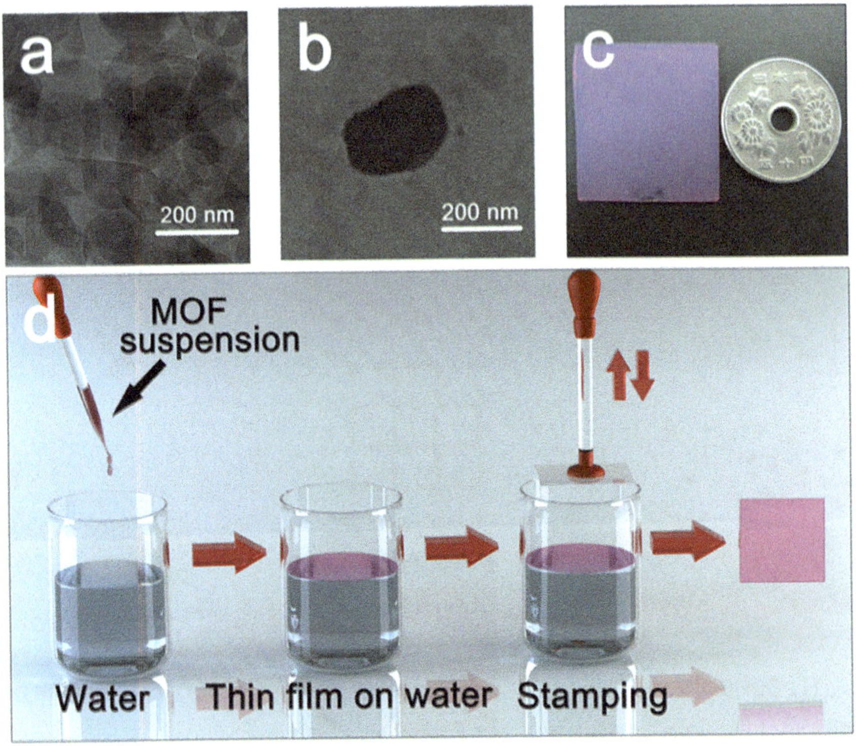

Fig. 1.5 **a**, **b** TEM images of Cu–TCPP nanosheets. **c** Photograph of the MOF thin film after 15 deposition cycles on a quartz substrate. **d** Illustration of the assembly process of this PCP thin film. Reprinted with permission from Ref. [36]. Copyright 2012 American Chemical Society

Thus, the doped PCPs are surrounded by a polymer matrix. In this case, the selectivity and permeability of the PCPs/polymer hybrid membrane are significantly dependent on the properties of the polymer. Liu et al. reported on the incorporation of engineered submicrometer-sized PCP crystals into polymers to form hybrid materials that successfully translate the excellent molecular sieving properties of face-centered cubic (fcu)-MOFs into the resultant membranes (Fig. 1.6) [37]. They demonstrate, simultaneously, exceptionally enhanced separation performance in hybrid membranes for two challenging and economically important applications: the removal of CO_2 and H_2S from natural gas and the separation of butane isomers.

The third type is self-supporting membrane. Self-supporting membrane is a kind of membrane without base support, which has certain mechanical strength and can withstand its own mass without breaking. The application of self-supporting film will not be limited by the substrate, so it has a wider application prospect. Currently, self-supported PCP membranes are mainly applied in gas separation, heavy metal filtration, ion conduction, and other fields [38]. Existing reports mainly focus on growing PCP crystals on fibers to enhance the film strength to self-supported [39],

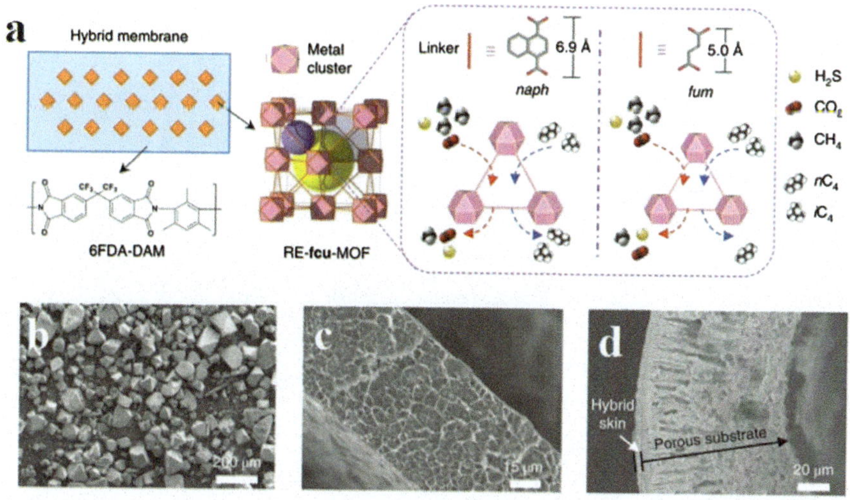

Fig. 1.6 Schematic of designing the MOF/polymer hybrid membrane (**a**). Scanning electron microscope images of as-synthesized Y-fum-fcu-MOF crystals (**b**). Cross-sectional hybrid membranes dense film with 17.8 wt% MOF loading (**c**) and asymmetric hollow fiber with ~13.0 wt% loading (**d**). Reprinted with permission from Ref. [37]. Copyright 2018 Springer Nature

growing self-supported films by two-phase interface method [40], and then removing the substrate after synthesis on the substrate [41]. However, for the films prepared by these methods, there are some problems such as no crystal orientation and the size and thickness of the films which cannot be accurately controlled. Peng et al. have reported a new method for preparing self-supporting membranes, which first synthesizes copper hydroxide nanowires through a simple mature chemical reaction and then filters them into thin films. Then, HKUST-1 with a thickness of 5 μm was successfully prepared by using the thin film as copper source in the reaction solvent of dissolving organic ligand homophonic acid. The self-supporting film is shown in Fig. 1.7 [42]. Meanwhile, the crystal growth process is monitored, and it is found that the crystal forming the film grows faster along the (001) direction than along the (111) direction. The results of gas separation experiments with the film show that the film has obvious selective permeability to hydrogen and can be used for hydrogen separation.

Pure PCP membranes should have the best performance in theory [43]. A key obstacle to the large-scale use of PCP-based materials in future practical applications is their poor machinability. PCPs are usually synthesized in the form of microcrystalline powders, which need to be shaped into various macroscopic bodies for easy handling and adaptation to current industrial processes. But their single crystals made of directional coordination bonds are inherently brittle, which severely limit their application [44]. Therefore, the development of a new method to prepare and synthesize highly ordered crystalline flexible PCP membranes is of great significance to promote the application of PCP materials.

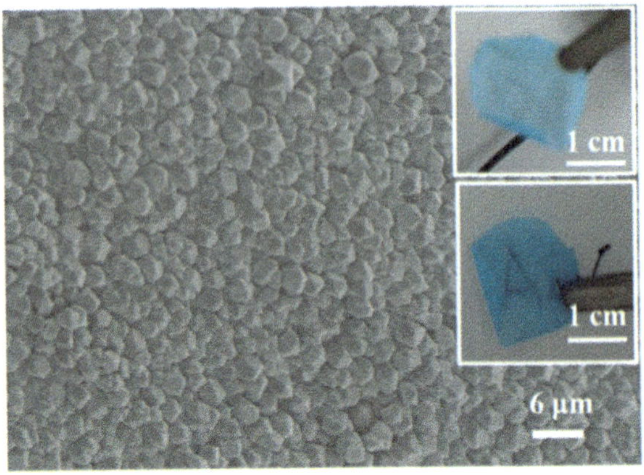

Fig. 1.7 Low magnification SEM image of the HKUST-1 membrane, and insets show that the membrane is transferable to another substrate and relatively transparent. Reprinted with permission from Ref. [42]. Copyright 2013 Royal Society of Chemistry

Two-dimensional (2D) PCP membranes assembled from ultra-thin PCP nanosheets have attracted great interest due to their ultra-thin film thickness, which provides a new strategy for the construction of high-performance membranes. The low structural dimension of CP often plays a key role in determining its flexibility, with the low-dimensional coordination chain acting as a rigid skeleton and being assembled along the other two directions by a large number of weak molecular interactions. In bending, the influence of coordination force is obviously reduced because the mechanical stress is mainly solved by weak bonding recombination. This provided an idea for us to prepare CPs' independent membranes. Two-dimensional (2D) CP membranes were formed by stacking ultra-thin CPs nanosheets through the weak interaction between the layers of the nanosheets.

1.3 Proton Rectification

Proton rectification (unidirectional proton transportation) is a key phenomenon for adenosine triphosphate production, which is driven by the pH gradient generated by proton pumps across cell membranes [45]. It is also critical for developing various artificial systems such as artificial photosynthesis [46], electrodialysis technology [47], and proton-exchange fuel cells [48]. However, implementation of the asymmetric proton transfer is not that easy, because at least a symmetry-breaking factor must be introduced into the proton pathway. However, introducing at least one asymmetric factor into the proton pathway (e.g., spatial structure, surface potential, and electrolyte media [49]) is the most serious obstacle to achieving the unidirectional proton transport.

A simple and practical example is the combination of cation- and anion-exchange membranes to impose a pH gradient, as applied in the first study on a liquid electrolyte-based proton rectifier device [50]. The device structure is best described by an electrolytic analogy with p–n diode, wherein protons and hydroxide ions are the dominant carrier species instead of electrons and holes. Specifically, an asymmetric surface potential is introduced to promote ion diffusion driven by the concentration gradient [51]. As subsequently discussed in detail, the pH gradient directs the diffusion current of protons/hydroxide ions toward the interfacial region, which is balanced by the drift current flowing in the opposite direction under no applied voltage (Fig. 1.8). Both proton and hydroxide ion diffusions are explicitly promoted and suppressed by applying forward and reverse bias voltages, which are primarily associated with the decrease and increase in the potential barrier (so-called built-in potential), respectively.

Compared to the liquid electrolyte-based proton rectifier devices, the fabrication of all-solid-state devices remains in the nascent stage despite their apparent advantages of thin configuration, high flexibility, and high safety against liquid leakage. The research of all-solid-state proton rectifiers is mainly through combining the cation- and anion-exchange membranes to impose a pH gradient to establish a barrier to achieve the rectification effect. In 1964, Seno et al. [52] built an all-solid-state rectifier by combining Amberlite IR-120 and Amberlite IRA-400 as cation- and anion-exchange membranes, respectively; the first reported all-solid-state rectifier. Figure 1.9 shows a schematic diagram of the rectifier. The rectification ratio is about 100, but considering that the rectification condition is that each membrane includes 56% weight of water content, that means the rectifier contains a lot of liquid water, not wholly belonging to an all-solid-state rectifier.

Langer first proposed the concept of H–OH junction similar to PN junction in 1978 [53]. They assumed that pure water was an intrinsic semiconductor and that

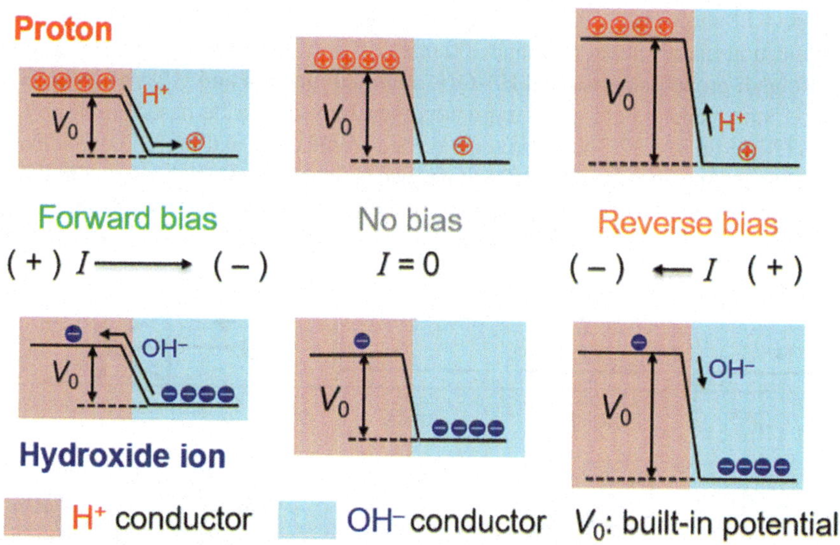

Fig. 1.8 Changes in potential (V_0) in H^+ and OH^- in a heterogeneous membrane by applying forward and reverse bias voltages

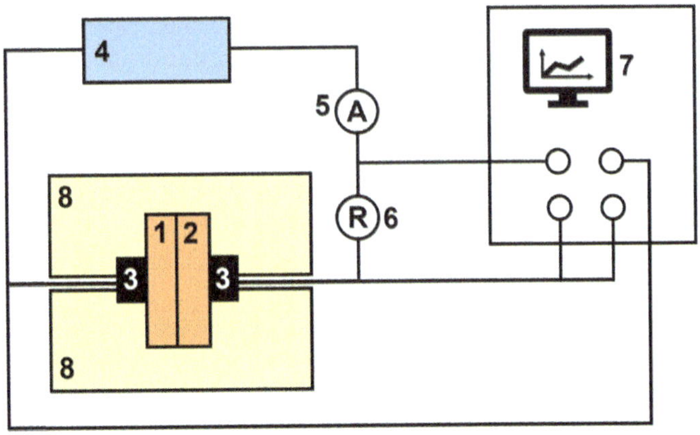

Fig. 1.9 Apparatus of rectifier. 1. Cation-selective membrane; 2. anion-selective membrane; 3. electrodes (stainless steel); 4. voltage source; 5. ammeter; 6. resistance (10 kΩ); 7. oscilloscope; 8. constant-moisture chamber [52]

water containing a small amount of acid or base was an N-type or P-type semiconductor, as shown in Fig. 1.10. He did this by binding Dowex-50 and Dowex-1 doped with cationic and anion exchangers, respectively. Dowex-50 and Dowex-1 were sandwiched by metallic electrodes, conditioned in high RH condition for 12 h, and then take rectification measurement. The *I–V* characteristic of our diode is very similar to the electronic diode. The forward direction of the protonic diode is just the same as predicted for protons as charge carriers. The breakdown voltage is 300 mV. Figure 1.11 shows the schematic diagram of the rectifier. Through calculation, the rectification ratio is about 6 around 300 mV.

Rolandi group demonstrated H^+–OH^- rectified junction and H^+-type and OH^--type complementary field-effect transistors using polysaccharide-based bionic proton lines [54]. Figure 1.12 Shows protonic device architecture and proton conductivity mechanism. Specifically, Fig. 1.12a shows two and three terminal devices with PdH_x source and drain. PdH_x is created by exposing Pd metal to 5% H_2 atmosphere. At this H_2 concentration, the Pd metal absorbs H_2 to form PdH_x with $x < 0.5$. PdH_x is kept

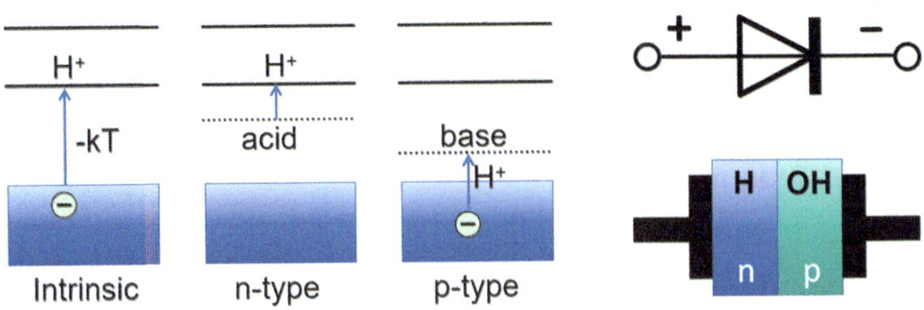

Fig. 1.10 Energy band scheme for protonic semiconductors [53]

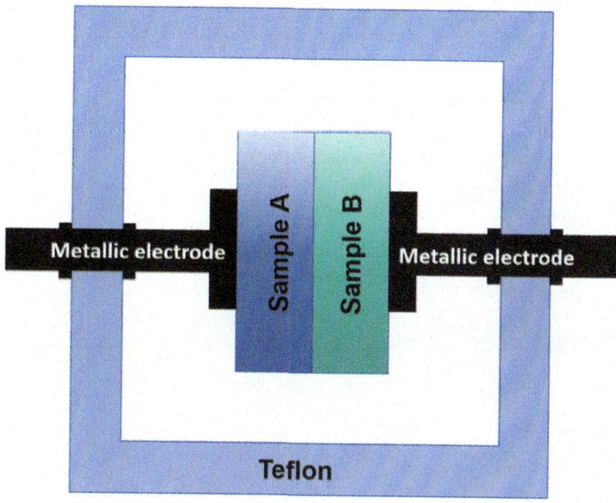

Fig. 1.11 Measuring cell used in this experiment [53]

under 5% H_2 atmosphere throughout the measurements and acts as a H^+ reservoir. The PdH_x source and drain inject and sink protons into and from the proton wire according to the reversible reaction PdH < PdH_x going from left to right at the source and from right to left at the drain. The PdH_x source and drain are connected to outside measurement electronics that measure the electronic current and complete the circuit. The rectification experiment was carried out 75% RH. The conductivity of OH^- as a proton hole is confirmed by proton wire. We describe the electrical conductivity in these devices by modeling that H^+ (proton) and OH^- (proton hole) are equivalent to electrons and holes in a semiconductor (Fig. 1.12c, d). Insights from the semiconductor model and protonic devices may prove useful at interrogating the conductivity in relevant proton channels from an alternate perspective.

Compared to the liquid electrolyte-based proton rectifier devices, the fabrication of all-solid-state devices remains in the nascent stage despite their apparent advantages of thin configuration, high flexibility, and high safety against liquid leakage.

1.4 The Structure of This Thesis

The purpose of this thesis is to report the fabrication of CP free-standing membranes and specific proton-conducting behavior in it. This section gives a short overview of the chapters and their interconnections.

Chapter 2, "**High-Performance All-Solid-State Proton Rectifier Using a Heterogeneous Membrane Composed of Coordination Polymer and Layered Double Hydroxide**," is focused on the fabrication of all-solid-state proton rectifier by the facile mechanical lamination of two free-standing membranes, namely, a proton-conducting PCP, $Cu_2(CuTCPP)$, and hydroxide ion-conducting LDH, Mg–Al–LDH(NO_3). The *I*–*V* characteristics provide a rectification ratio as high as > 200 at 25 °C under 90% RH, which is the highest recorded value.

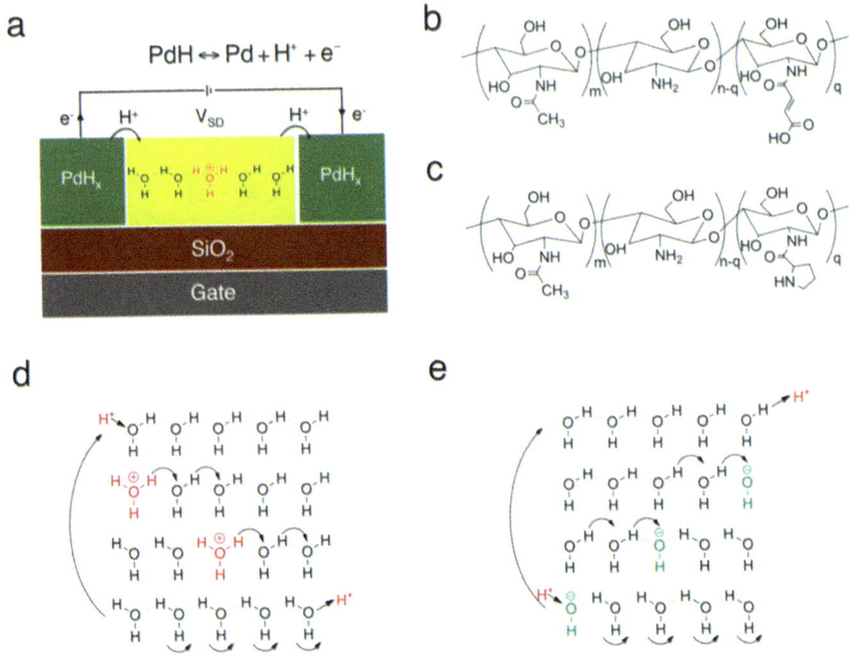

Fig. 1.12 Proton device design and transfer mechanisms. **a** PdH_x-based devices. **b** H^+ conductor: maleic chitosan. **c** OH^- conductor: proline chitosan. Substitution ratio $q/(n + m)$ affects doping. **d** H^+ transport via Grotthuss mechanism. **e** OH^- transport via similar mechanism. Reprinted with permission from Ref. [54]. Copyright 2013 Springer Nature

In Chap. 3, "**Robust Proton Conduction against Mechanical Stress in Flexible Free-Standing Membrane Composed of Two-Dimensional Coordination Polymer**," we fabricated highly flexible free-standing membranes composed of 2D CP, Cu_2(NiTCPP). The membrane is highly oriented and excellent mechanical properties by virtue of weak interlayer interactions of van der Waals type. The proton conductivity of the membrane remains almost unchanged even by applying bending stress. The proton-conducting properties, as well as the structural features, in different bending states strongly indicate that proton-conducting pathway through the hydrogen-bonding network is kept intact during the bending operation. This is the first study to demonstrate the robust proton conduction of 2D CP against the applied bending stress.

Following Chap. 4, "**Proton Conducting Behavior in Flexible Coordination Polymer Free-Standing Membranes**," we fabricated highly flexible free-standing membranes composed of 2D CPs, Cu_2(NiTCPP), Cu_2(CuTCPP), and Cu_2(ZnTCPP). The membrane was highly oriented and had excellent mechanical properties owing to the weak interlayer interactions of van der Waals type. The σ value of the membrane reached 4.53×10^{-5}, 2.00×10^{-5}, and 4.73×10^{-5} S cm^{-1} at 45 °C and 98% RH condition. This is the first study to fabricate a series of flexible CP free-standing membranes. The prominent advantage of the present membrane for future flexible protonic devices is that it can be fabricated free from any substrate and shows excellent bending stability even without blending with the polymer matrix.

Finally, this Ph.D. thesis is concluded with general conclusion in Chap. 5.

References

1. (a) O.M. Yaghi, M. O'Keeffe, N.W. Ockwig, H.K. Chae, M. Eddaoudi, J. Kim, Nature **423,** 705–714 (2003); (b) S. Kitagawa, R. Kitaura, S.I. Noro, Angew. Chem. Int. Ed. **43,** 2334–2375 (2004), Angew. Chem. 2004, 43, 2334–2375; (c) H.C. Zhou, J.R. Long, O. M. Yaghi, Chem. Rev. **112,** 673–674 (2012); (d) H. Furukawa, K.E. Cordova, M. O'Keeffe, O.M. Yaghi, Science **341,** 1230444 (2013); (e) S. Kitagawa, Chem. Soc. Rev. **43,** 5415–5418 (2014)
2. (a) J. Lee, O.K. Farha, J. Roberts, K.A. Scheidt, S.T. Nguyen, J.T. Hupp, Chem. Soc. Rev. **38,** 1450–1459 (2009); (b) Y.B. Huang, J. Liang, X.S. Wang, R. Cao, Chem. Soc. Rev. **46,** 126–157 (2017); (c) J. Liu, L. Chen, H. Cui, J. Zhang, L. Zhang, C.Y. Su, Chem. Soc. Rev. **43,** 6011–6061 (2014)
3. (a) E.A. Dolgopolova, A.M. Rice, C.R. Martin, N.B. Shustova, Chem. Soc. Rev. **47,** 4710–4728 (2018); (b) Y. Liu, X.Y. Xie, C. Cheng, Z.S. Shao, H.S. Wang, J. Mater. Chem. C **7,** 10743–10763 (2019); (c) M. Daniel, G. Mathew, M. Anpo, B. Neppolian, Coord. Chem. Rev. **468,** 214627 (2022)
4. (a) H. Li, L. Li, R.B. Lin, W. Zhou, Z. Zhang, S. Xiang, B. Chen, Energy Chem **1,** 100006 (2019); (b) W. Fan, X. Zhang, Z. Kang, X. Liu, D. Sun, Coord. Chem. Rev. **443,** 213968 (2021); (c) B. Li, H.M. Wen, Y. Yu, Y. Cui, W. Zhou, B. Chen, G. Qian, Mater. Today Nano **2,** 21–49 (2018); (d) D.X. Xue, Q. Wang, J. Bai, Coord. Chem. Rev. **378,** 2–16 (2019); (e) H. Li, K. Wang, Y. Sun, C.T. Lollar, J. Li, H.C. Zhou, Mater. Today **21,** 108–121 (2018); (f) K.A. Cychosz, M. Thommes, Engineering **4,** 559–566 (2018)
5. J. Canivet, A. Fateeva, Y. Guo, B. Coasne, D. Farrusseng, Chem. Soc. Rev. **43**, 5594–5617 (2014)
6. (a) Y. Ye, L. Gong, S. Xiang, Z. Zhang, B. Chen, Adv. Mater. **32,** 1907090 (2020); (b) S. Paul, S. J. Choi, H.J. Kim, Energy Fuels **34,** 10067–10077 (2020)
7. G.D. Wu, H.L. Zhou, Z.H. Fu, W.H. Li, J.W. Xiu, M.S. Yao, Q.H. Li, G. Xu, Angew. Chem. Int. Ed. **60,** 9931–9935 (2021), Angew. Chem. 2021, 133, 10019-10023
8. (a) S.Z. Huang, S.S. Liu, H.J. Zhang, Z. Han, G. Zhao, X.Y. Dong, S.Q. Zang, ACS Appl. Mater. Interfaces **12,** 28720–28726 (2020); (b) L.D. Costanzo, B. Panunzi, Molecules **26,** 2952 (2021); (c) S.L. Yang, G. Li, M.Y. Guo, W.S. Liu, R. Bu, E. Q. Gao, J. Am. Chem. Soc. **143,** 8838–8848 (2021)
9. (a) X. Li, H. Zhang, J. Hou, R. Ou, Y. Zhu, C. Zhao, T.Y. Qian, C.D. Easton, C. Selomulya, M.R. Hill, H. Wang, J. Am. Chem. Soc. **142,** 9827–9833 (2020); (b) J. Lu, H. Xu, H. Yu, X. Hu, J. Xia, Y. Zhu, F.C. Wang, H.A. Wu, L. Jiang, H. Wang, Sci. Adv. **8,** eabl5070 (2022)
10. S. Kanda, K. Yamashita, K. Ohkawa, Bull. Chem. Soc. Jpn **52**, 3296 (1979)
11. (a) G.K.H. Shimizu, J.M. Taylor, S. Kim, Science **341,** 354 (2013); (b) T. Yamada, K. Otsubo, R. Makiura, H. Kitagawa, Chem. Soc. Rev. **42,** 6655–6669 (2013); (c) D.W. Lim, H. Kitagawa, Chem. Rev. **120,** 8416–8467 (2020); (d) D.W. Lim, M. Sadakiyo, H. Kitagawa, Chem. Sci. **10,** 16–33 (2019)
12. (a) N. Agmon, Chem. Phys. Lett. **244,** 456 (1995); (b) K.D. Kreuer, A. Rabenau, W. Weppner, Angew. Chem. Int. Ed. **21,** 208 (1982)
13. T. Ueki, M. Watanabe, Macromolecules **41**, 3739–3749 (2008)
14. J. Zhang, H.-J. Bai, Q. Ren, H.-B. Luo, X.-M. Ren, Z.-F. Tian, S. Lu, A.C.S. Appl, Mater. Interfaces **10**, 28656 (2018)
15. J. Jaafar, M. Nordin, N.A. Hadi, A.F. Ismail, M.H.D. Othman, M.A. Rahman, F. Aziz, J. Membr. Sci. Res. **5**, 65 (2019)

16. (a) B.D. Cahan, J.S. Wainright, J. Electrochem. Soc. **140,** L185 (1993); (b) Y. Sone, P. Ekdunge, D. Simonsson, J. Electrochem. Soc. **143**(4), 1254 (1996); (c) C. Ma, L. Zhang, S. Mukerjee, D. Ofer, B. Nair, J. Membr. Sci. **219,** 123–136 (2003)
17. D. Umeyama, S. Horike, M. Inukai, T. Itakura, S. Kitagawa, J. Am. Chem. Soc. **134**, 12780 (2012)
18. M. Liu, L. Chen, S. Lewis, S.Y. Chong, M.A. Little, T. Hasell, I.M. Aldous, C.M. Brown, M.W. Smith, C.A. Morrison, L.J. Hard- Wick, A.I. Cooper, Nat. Commun. **7,** 12750 (2016)
19. K. Zhang, X. Xie, H. Li, J. Gao, L. Nie, Y. Pan, J. Xie, D. Tian, W. Liu, Q. Fan, H. Su, L. Huang, W. Huang, Adv. Mater. **29**, 1701804 (2017)
20. L. Wu, Y. Yang, Y. Ye, Z. Yu, Z. Song, S. Chen, L. Chen, Z. Zhang, S. Xiang, ACS Appl. Energy Mater. **1**, 5068 (2018)
21. F. Yang, H. Huang, X. Wang, F. Li, Y. Gong, C. Zhong, J.-R. Li, Cryst. Growth Des. **15**, 5827 (2015)
22. P. Losch, H.R. Joshi, O. Vozniuk, A. Grünert, C. Ochoa-Hernández, H. Jabraoui, M. Badawi, W. Schmidt, J. Am. Chem. Soc. **140**, 17790 (2018)
23. X. Shi, Y. Shan, M. Du, H. Pang, Coord. Chem. Rev. **444**, 214060 (2021)
24. I. Hod, M.D. Sampson, P. Deria, C.P. Kubiak, O.K. Farha, J.T. Hupp, ACS Catal. **5**, 6302–6309 (2015)
25. S. Sakaida, K. Otsubo, O. Sakata, C. Song, A. Fujiwara, M. Takata, H. Kitagawa, Nat. Chem. **8**, 377–383 (2016)
26. S. Hurrle, S. Friebe, J. Wohlgemuth, C. Wöll, J. Caro, L. Heinke, Chem. Eur. J. **23**, 2294–2298 (2017)
27. J. Liu, W. Zhou, J. Liu, I. Howard, G. Kilibarda, S. Schlabach, D. Coupry, M. Addicoat, S. Yoneda, Y. Tsutsui, T. Sakurai, S. Seki, Z. Wang, P. Lindemann, E. Redel, T. Heine, C. Wöll, Angew. Chem. Int. Ed. **54**, 7441–7445 (2015)
28. X. Shen, B. Yan, Dalt. Trans. **44**, 1875–1881 (2015)
29. Y.J. Jang, Y.E. Jung, G.W. Kim, C.Y. Lee, Y.D. Park, RSC Adv. **9**, 529–535 (2019)
30. J.J. Richardson, M. Bjornmalm, F. Caruso, Science **348**, 6233 (2015)
31. Y. Liu, Z. Ng, E.A. Khan, H.K. Jeong, C. Ching, Z. Lai, Microporous Mesoporous Mater. **118**, 296–301 (2009)
32. R. Makiura, R. Usui, Y. Sakai, A. Nomoto, A. Ogawa, O. Sakata, A. Fujiwara, ChemPlusChem **79**, 1352–1360 (2014)
33. J. Cho, K. Char, J.D. Hong, K.B. Lee, Adv. Mater. **13**, 1076–1078 (2001)
34. M.C. So, S. Jin, H.J. Son, G.P. Wiederrecht, O.K. Farha, J.T. Hupp, J. Am. Chem. Soc. **135**, 15698–15701 (2013)
35. Y. Hu, X. Dong, J. Nan, W. Jin, X. Ren, N. Xu, Y.M. Lee, ChemComm. **47**, 737–739 (2011)
36. G. Xu, T. Yamada, K. Otsubo, S. Sakaida, H. Kitagawa, J. Am. Chem. Soc. **134**, 16524–16527 (2012)
37. G. Liu, V. Chernikova, Y. Liu, K. Zhang, Y. Belmabkhout, O. Shekhah, C. Zhang, S. Yi, M. Eddaoudi, W.J. Koros, Nat. Mater. **17**, 283–289 (2018)
38. (a) R. Ameloot, F. Vermoortele, W. Vanhove, M.B. Roeffaers, B.F. Sels, D.E. De Vos, Nat. Chem. **3,** 382–387 (2011); (b) Y. Zhang, X. Feng, H. Li, v Chen, J. Zhao, S. Wang, L. Wang, B. Wang, Angew. Chem. Int. Ed. **54,** 4259–4263 (2015); (c) S. Qiu, M. Xue, G. Zhu, Chem. Soc. Rev. **43,** 6116–6140 (2014)
39. Y.N. Wu, F. Li, H. Liu, W. Zhu, M. Teng, Y. Jiang, W. Li, D. Xu, D. He, P. Hannam, G. Li, J. Mater. Chem. **22**, 16971–16978 (2012)
40. Z. Zeng, Z. Yin, X. Huang, H. Li, Q. He, G. Lu, F. Boey, H. Zhang, Angew. Chem. **123**, 11289–11293 (2011)
41. M. Darbandi, H. K. Arslan, O. Shekhah, A. Bashir, A. Birkner, C. Wöll, Phys. Status Solidi (RRL) Rapid Res. Lett. **4,** 197–199 (2010)
42. Y. Mao, H. Huang, W. Cao, J. Li, L. Sun, X. Jin, X. Peng, ChemComm. **49**, 5666–5668 (2013)
43. S. Jia, D. Ji, L. Wang, X. Qin, S. Ramakrishna, Small Struct. **3**, 2100222 (2022)
44. L.C. An, X. Li, Z.G. Li, Q. Li, P.J. Beldon, F.F. Gao, Z.Y. Li, S. Zhu, L. Di, S. Zhao, J. Zhu, D. Comboni, I. Kupenko, W. Li, U. Ramamurty, X.H. Bu, Nat. Commun. **13**, 6645 (2022)

45. (a) H.C. Blair, S.L. Teitelbaum, R. Ghiselli, S. Gluck, Science **245,** 855–857 (1989); (b) G. Steinberg-Yfrach, J.L. Rigaud, E.N. Durantini, A.L. Moore, D. Gust, T.A. Moore, Nature **392,** 479–482 (1998); (c) D. Gadsby, Nat. Rev. Mol. Cell Biol. **10,** 344–352 (2009)
46. (a) D. Gust, T.A. Moore, A.L. Moore, Acc. Chem. Res. **34,** 40–48 (2001); (b) P.K. Ghosh, A.Y. Smirnov, F. Nori, J. Chem. Phys. **131,** 035102 (2009); (c) S. Berardi, S. Drouet, L. Francàs, C. Gimbert-Suriñach, M. Guttentag, C. Richmond, T. Stoll, A. Llobet, Chem. Soc. Rev. **43,** 7501–7519 (2014); (d) S. Chabi, K.M. Papadantonakis, N.S. Lewis, M.S. Freund, Energy Environ. Sci. **10,** 1320–1338 (2017)
47. (a) Y. Tanaka, Membr. Sci. Tech. **12,** 405–436 (2007); (b) R. Pärnamäe, S. Mareev, V. Nikonenko, S. Melnikov, N. Sheldeshov, V. Zabolotskii, H.V.M. Hamelers, M. Tedesco, J. Membr. Sci. **617,** 118538 (2021)
48. K. Jiao, J. Xuan, Q. Du, Z. Bao, B. Xie, B. Wang, Y. Zhao, L. Fan, H. Wang, Z. Hou, S. Huo, N.P. Brandon, Y. Yin, M.D. Guiver, Nature **595**, 361–369 (2021)
49. W. Guo, Y. Tian, L. Jiang, Acc. Chem. Res. **46**, 2834–2846 (2013)
50. B. Lovrecek, A. Despic, J.O.M. Bockris, J. Phys. Chem. **63**, 750–751 (1959)
51. (a) P.K. Giesbrecht, M.S. Freund, Chem. Mater. **32,** 8060–8090 (2020); (b) Z. Zhang, X. Y. Kong, K. Xiao, Q. Liu, G. Xie, P. Li, J. Ma, Y. Tian, L.P. Wen, L. Jiang, J. Am. Chem. Soc. **137,** 14765–14772 (2015)
52. M. Seno, T. Yamabe, Bull. Chem. Soc. Jpn **37**, 668–671 (1964)
53. J.J. Langer, Appl. Phys. A **34**, 195–198 (1984)
54. Y. Deng, E. Josberger, J. Jin, A.F. Roudsari, B.A. Helms, C. Zhong, M.P. Anantram, M. Rolandi, Sci. Rep. **3**, 2481 (2013)

Chapter 2
High-Performance All-Solid-State Proton Rectifier Using a Heterogeneous Membrane Composed of Coordination Polymer and Layered Double Hydroxide

Abstract Rational control of unidirectional proton transport is highly challenging, primarily owing to the difficulty in introducing an asymmetric factor into proton-conducting media. In this study, free-standing membranes of a proton-conducting two-dimensional porous coordination polymer, $Cu_2(CuTCPP)$ (H_2TCPP: 5,10,15,20-tetrakis(4-carboxyphenyl)porphyrin) and a hydroxide ion-conducting layered double hydroxide, Mg–Al–LDH(NO_3), were combined to generate a pH gradient in the conducting media. The current–voltage measurements revealed that the heterogeneous membrane exhibits a significant unidirectional proton transport with a proton rectification ratio exceeding 200 under 90% relative humidity in the initial voltage scan. This value is the highest among the reported all-solid-state proton rectifiers. The high designability of both components with well-defined structures, which is in contrast to the organic polymers used so far, provides a new avenue for developing and understanding the proton-rectifying behavior in the solid state.

Keywords All-solid-state · Proton transport · Coordination polymer · Rectification

2.1 Introduction

Unidirectional proton transport plays an essential role in adenosine triphosphate production, which is driven by the pH gradient generated by proton pumps across cell membranes [1]. It is also critical for developing various artificial systems such as artificial photosynthesis [2], electrodialysis technology [3], and proton-exchange fuel cells [4]. However, introducing at least one asymmetric factor into the proton pathway (e.g., spatial structure, surface potential, and electrolyte media [5]) is the most serious obstacle to achieving the unidirectional proton transport. A simple and practical example is the combination of cation- and anion-exchange membranes to impose a pH gradient, as applied in the first study on a liquid electrolyte-based proton rectifier device [6]. The device structure is best described by an electrolytic analogy with p–n diode, wherein protons and hydroxide ions are the dominant carrier species instead

J. Lu, *Novel Proton Conducting in Free-Standing Coordination Polymer Membranes*, Springer Theses, https://doi.org/10.1007/978-981-97-7913-0_2

of electrons and holes. Specifically, an asymmetric surface potential is introduced to promote ion diffusion driven by the concentration gradient [7]. As subsequently discussed in detail, the pH gradient directs the diffusion current of protons/hydroxide ions toward the interfacial region, which is balanced by the drift current flowing in the opposite direction under no applied voltage (Fig. 2.1). Application of a forward bias voltage explicitly promotes both proton and hydroxide ion diffusions, which is primarily associated with the decrease in the potential barrier (so-called built-in potential). In contrast, the reverse bias voltage suppresses the diffusions due to the increased potential barrier.

Compared to the liquid electrolyte-based proton rectifier devices [5, 8], the fabrication of all-solid-state devices remains in the nascent stage [9] despite their apparent advantages of thin configuration, high flexibility, and high safety against liquid leakage. In addition, the components of membranes in all-solid-state proton rectifiers have been limited to amorphous organic polymers with low designability, leading to the difficulty in studying the rectifying behavior from a structural perspective. To overcome these drawbacks, we adopted structurally defined porous coordination polymers (PCPs) [10] as proton conductors [11]. The high designability of PCPs, which primarily originates from their diverse combinations of metal nodes and organic linkers, enables rational control of porous structures with tunable sizes, topologies, and surface potentials [12]. In particular, PCPs with high proton conduction have been extensively investigated by judicious

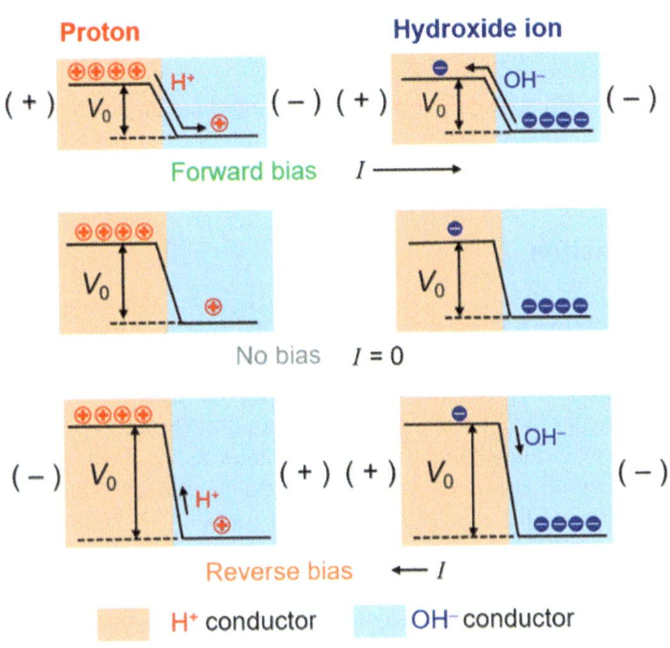

Fig. 2.1 Changes in potential (V_0) in H^+ and OH^- in a heterogeneous membrane by applying forward and reverse bias voltages

selection of the components, wherein water molecules included in the pores serve as proton hopping sites [13]. However, unidirectional proton transport has been largely unexplored in PCP systems; to date, no PCP-based all-solid-state proton rectifiers capable of preventing liquid leakage have been fabricated. We selected $Cu_2(CuTCPP)$ ($H_4(H_2TCPP)$; 5,10,15,20-tetrakis(4-carboxyphenyl)porphyrin) with a two-dimensional (2D) square grid structure composed of copper porphyrins and paddlewheel-type copper dimers (Fig. 2.2) [14], because it exhibits a high in-plane proton conductivity (3.9×10^{-3} S cm^{-1} at 25 °C under 98% relative humidity (RH)) and can be assembled into 2D nanosheets [15]. The moderate intersheet interaction mainly through van der Waals force provides considerable freedom in the stacking manner, which favors the formation of a free-standing membrane [16]. As a hydroxide ion conductor, Mg–Al-layered double hydroxide (LDH), with 2D layers composed of edge-shared Mg(II) and Al(III) hydroxide octahedra (Fig. 2.3) [17], was selected, because the in-plane hydroxide ion conductivity (1×10^{-1} S cm^{-1} at 60 °C under 80% RH) is among the highest reported to date, and the 2D layers can be readily exfoliated into nanosheets, as in $Cu_2(CuTCPP)$ [18]. Theoretical calculations predicted that guest water molecules energetically prefer to transfer protons to immobile hydroxyl sites in Mg–Al–LDH layers to generate mobile OH^- ions ($M(OH)_2 + xH_2O \leftrightarrow M(H_2O)_x(OH)_{2-x}{}^{x+} + xOH^-$; Fig. 2.4) [19]. In this study, free-standing membranes of each ion conductor were separately prepared, followed by the mechanical fabrication of their heterogeneous membrane, which was sandwiched by Pd electrodes. The electrodes can serve as proton transmission media and catalysts for water dissociation, as discussed subsequently. Particularly, both components with layered structures can adsorb a considerable amount of water, which facilitates the integration of solvent-free rectifier devices without the risk of liquid leakage. Our device demonstrated an ion rectification rate (i.e., the ratio of the forward/reverse bias currents) greater than 200, which is the highest among the reported all-solid-state H^+–OH^- rectifiers.

2.2 Experimental

2.2.1 Materials

Copper(II) nitrate trihydrate (99%), magnesium(II) nitrate hexahydrate (99%), aluminum(III) nitrate nonahydrate (98%), sodium nitrate (99%), nitric acid (60–61 wt.%), hexamethylenetetramine (99%), *N*,*N*-dimethylformamide (DMF, 99.5%), and potassium bromide (crystal block) were used as received from Fujifilm Wako Pure Chemical Industries, Ltd. Tetrakis(4-carboxyphenyl)porphyrin ($H_4(H_2TCPP)$, 97%) was purchased from Tokyo Chemical Industry Co., Ltd., and used without further purification. Water was deionized before use.

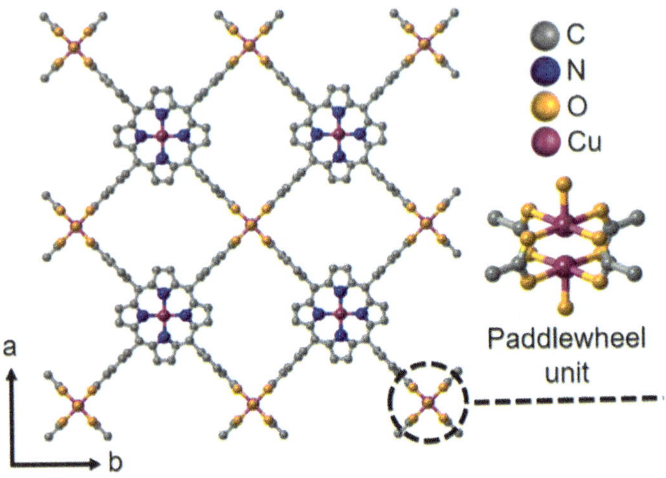

Fig. 2.2 2D sheet structure of $Cu_2(CuTCPP)$

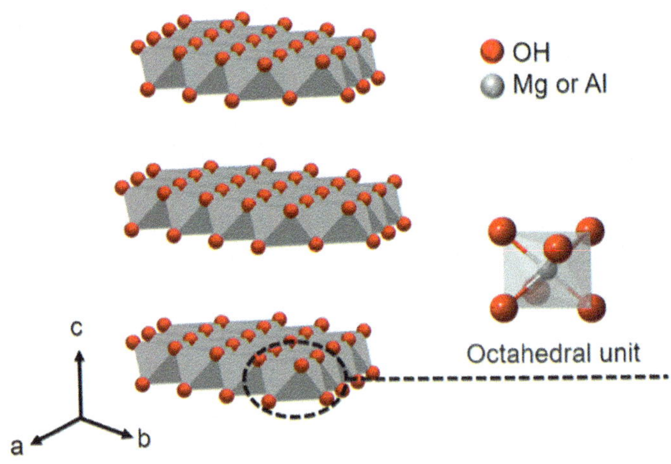

Fig. 2.3 Layered structure of Mg–Al–LDH(NO_3)

2.2.2 *Characterizations*

Powder X-ray diffraction (PXRD) analysis was performed using a Bruker D8 ADVANCE with Cu K_α radiation ($\lambda = 0.15418$ nm) in the 2θ range from 5° to 60° with a 0.02° step and a step time of 1 s. In-plane and out-of-plane XRD scans were measured using a Rigaku Smart-Lab X-ray diffractometer with Cu K_α radiation ($\lambda =$ 0.15418 nm) in the 2θ range from 5 to 60° at a scanning rate of 0.5° min^{-1}. Fourier-transform infrared (FTIR) spectra were obtained using a PerkinElmer Spectrum 100 spectrometer. UV–vis absorption spectra were recorded on KBr pellets using a

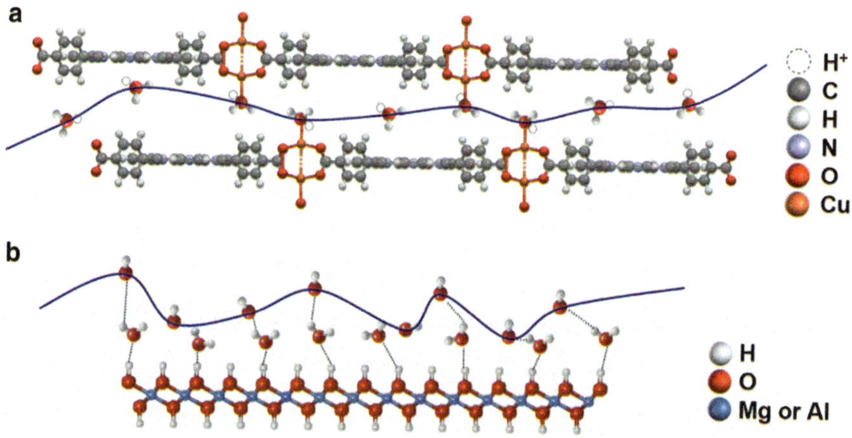

Fig. 2.4 Schematic illustrations of the proposed pathway and mechanism of **a** proton transport in Cu_2(CuTCPP) and **b** hydroxide ion transport in Mg–Al–LDH(NO_3), where blue lines indicate the proposed conducting pathways

PerkinElmer Lambda 1050 UV/Vis/NIR Spectrometer. Thermal stability was characterized by thermogravimetric (TG) measurements using a Bruker TG-DTA2000SA instrument at a heating rate of 5 °C min^{-1} under flowing N_2. Water vapor adsorption isotherms were measured at 298 K using a BELSORP-max volumetric adsorption system. Scanning electron micrographs (SEM) were obtained using a Hitachi SU-1510 instrument. Transmission electron microscope (TEM) images were obtained using a Hitachi HT7700 instrument. The impedance spectra of the membranes were measured using an Impedance/Gain-Phase Analyzer SI 1260 at an applied a.c. voltage of 100 mV. X-ray photoelectron spectroscopy (XPS) data were obtained with a Shimadzu ESCA-3400 X-ray photoelectron spectrometer. X-ray fluorescence (XRF) analysis was performed with a Rigaku ZSX Primus IV instrument.

2.2.3 *Preparation of Cu_2(CuTCPP) Nanosheets*

Cu_2(CuTCPP) nanosheets were synthesized according to a previous procedure [15]. Typically, 36.5 mg (0.15 mmol) of $Cu(NO_3)_2 \cdot 3H_2O$ was dispersed in 40 mL of DMF, whereas 40.0 mg (0.05 mmol) of H_2(H_4TCPP) was dissolved into 40 mL of DMF. These solutions were mixed and ultrasonicated for 10 min. The mixture was aged at 35 °C for 10 h and then heated at 85 °C for 24 h. The mixture was cooled to room temperature within 12 h. Finally, the gel product was collected by centrifugation (7830 rpm for 10 min) and washed with ethanol more than five times until the supernatant became colorless. The purple polycrystalline product was then dried at 50 °C for 12 h (yield: approximately 80%).

2.2.4 Preparation of Mg–Al–LDH(NO_3) Nanosheets

A CO_3-containing precursor, Mg–Al-LDH(CO_3), was synthesized according to a previous procedure [18]. Typically, 5.74 g (0.02 mol) of $Mg(NO_3)_2 \cdot 6H_2O$, 3.75 g (0.01 mol) of $Al(NO_3)_3 \cdot 9H_2O$, and 3.64 g (0.026 mol) of hexamethylenetetramine were dissolved in 80 mL of water. The mixture was placed in a stainless steel autoclave and heated at 140 °C for 24 h. The white polycrystalline product was recovered by filtration, washed with water, and air-dried (yield: approximately 70%).

Polycrystalline Mg–Al–LDH(CO_3) (0.5 g) was further treated with 500 mL of water containing $NaNO_3$ (0.75 mol) and HNO_3 (0.0025 mol) to expel interlayer carbonate ions. After purging with nitrogen gas, the glass flask was tightly sealed and shaken for 24 h at room temperature. The resulting product was collected by centrifugation (7830 rpm for 5 min) and washed with water more than five times until the supernatant became colorless. Finally, the white polycrystalline product was dried at 60 °C for 12 h (yield: approximately 40%). Mg/Al = 2.00 by XRF.

2.2.5 Preparation of Cu_2(CuTCPP) Free-Standing Membrane

Cu_2(CuTCPP) free-standing membrane was fabricated according to a previous procedure [15]. Forty mg of polycrystalline Cu_2(CuTCPP) was added to 200 mL of ethanol in a 300 mL flask, which was tightly sealed after purging with nitrogen gas. Ultrasonic treatment for 15 min allowed the exfoliated nanosheets to disperse well in ethanol. A Cu_2(CuTCPP) free-standing membrane was prepared by filtering the ethanol solution in which the Cu_2(CuTCPP) nanosheets were dispersed. Briefly, 20 mL of the ethanolic Cu_2(CuTCPP) suspension (0.2 mg mL^{-1}) was vacuum-filtered through a polyvinylidene difluoride (PVDF) membrane filter (pore diameter: ~0.22 μm; porosity: ~80%; diameter: 25 mm) and dried at 50 °C for 4 h. The Cu_2(CuTCPP) membrane, with a thickness of approximately 20 μm, could be easily detached from the filter to allow free-standing operation.

2.2.6 Preparation of Mg–Al–LDH(NO_3) Free-Standing Membrane

Mg–Al–LDH(NO_3) free-standing membrane was fabricated according to a previous procedure [18]. Fifty mg of polycrystalline Mg–Al–LDH(NO_3) was added to 100 mL of formamide in a 200 mL flask, which was tightly sealed after purging with nitrogen gas. Ultrasonic treatment for 3 h allowed the exfoliated nanosheets to disperse well in formamide. A Mg–Al–LDH(NO_3) free-standing membrane was prepared by filtering the formamide solution in which the Mg–Al–LDH(NO_3) nanosheets were dispersed. Briefly, 20 mL of the formamide Mg–Al–LDH(NO_3) suspension (0.5 mg mL^{-1}) was vacuum-filtered through a PVDF membrane filter (pore diameter: ~0.22 μm; porosity: ~80%; diameter: 25 mm) and dried at 60 °C for 20 h. The Mg–Al–LDH(NO_3) membrane, with a thickness of approximately 20 μm, could be easily detached from the filter to allow free-standing operation.

2.2.7 *Fabrication of an All-Solid-State Rectifier Device*

A piece of Pd (lateral size: 6 × 4 mm^2, thickness: 0.1 mm) was welded onto an SUS403 plate (lateral size: 16 × 4 mm^2, thickness: 0.8 mm) using a spot-welding device (KTH-MWS, Kondo Technology); the electrode was polished using 2000 mesh sandpaper. Free-standing membranes of Cu_2(CuTCPP) and Mg–Al–LDH(NO_3) with a thickness of approximately 20 μm were sandwiched between two Pd electrodes with physical force (2.0(1) N) to form the all-solid-state rectifier device.

2.2.8 *Rectification Behavior Measurements*

Current–voltage (*I–V*) curves were measured using a Model 4200-SCS Semiconductor Characterization System (Keithley Instruments Corp.), whereas time-dependent current curves were measured using a 6487 Picoammeter/Voltage Source (Keithley Instruments Corp.). In the *I–V* measurements, a reverse bias scan (0 to −10 V) was performed before a forward bias scan (0–10 V) to prevent water formation under forwarding bias from influencing the reverse bias test. Each data point was recorded after a waiting time of 0.1 s to eliminate transient effects. Rectification performance was tested in a temperature- and humidity-controlled chamber (ESPEC SH-221). Before testing, the heterogeneous membrane was placed in the chamber for more than 4 h to attain a stable state.

2.3 Results and Discussion

Purple crystalline Cu_2(CuTCPP) nanosheets were obtained through a solvothermal reaction as reported previously (see Experimental part for details) [15c]. The infrared absorption spectrum showed the disappearance of the C=O stretching vibration in COOH groups at 1676 cm^{-1} (Fig. 2.5), indicating that the four carboxyl groups of CuTCPP successfully coordinate to adjacent copper ions to form dimeric copper paddlewheel units [14]. Sheet-like structure is evident in the scanning electron microscope (SEM) image with a lateral dimension of approximately 3 μm (Fig. 2.6a). The transmission electron microscope (TEM) image of Cu_2(CuTCPP) deposited by dropping the ethanol suspension onto a carbon-coated copper substrate exhibited a nanosheet-like morphology (Fig. 2.6b). Because the Cu_2(CuTCPP) nanosheets are well dispersed in ethanol, the ethanol suspension was filtered through a polyvinylidene difluoride (PVDF) membrane filter to yield a highly oriented membrane (see Experimental part for details). The membrane with good flexibility (Fig. 2.7b) was readily peeled from the filter after drying at 50 °C (Fig. 2.7a). Thicknesses ranging from tens to hundreds of micrometers can be controlled using different amounts of Cu_2(CuTCPP) suspension. The cross-sectional SEM image (Fig. 2.8a) showed that the membrane is formed by stacking the nanosheets, which is consistent with the SEM image of the membrane surface (Fig. 2.8b). Figure 2.9 shows the out-of-plane and in-plane X-ray diffraction (XRD) patterns of the membrane. In contrast to the in-plane pattern, the out-of-plane pattern is dominated by an interlayer (002) reflection, indicating the preferred orientation of the membrane along the crystallographic *c* axis. The interlayer distance of 0.45 nm is approximately similar to that reported previously [15a, 15c].

Similarly, a white membrane of Mg–Al–LDH(NO_3) (NO_3^- as a counterion in the interlayer space) was obtained by filtering the formamide suspension of the nanosheets with a regular hexagonal platelet morphology (lateral dimension of approximately 4 μm) [21], as observed by SEM (Fig. 2.10a) and TEM (Fig. 2.10b) images (see Experimental part for details). The membrane, which also exhibited good flexibility (Fig. 2.11b), was readily peeled off from the PVDF filter after heating at 60 °C (Fig. 2.11a). The cross-sectional SEM image confirmed that the nanosheets

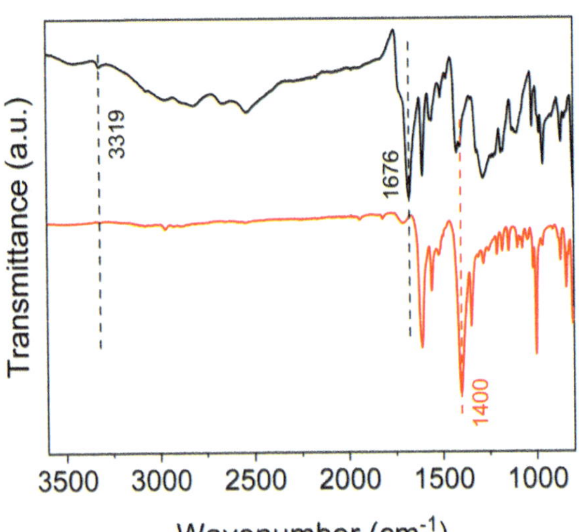

Fig. 2.5 Fourier-transform infrared spectra of $H_4(H_2TCPP)$ (black line) and Cu_2(CuTCPP) (red line). C=O stretching (1676 cm^{-1}) and N–H stretching (3319 cm^{-1}) bands, which are found in $H_4(H_2TCPP)$, disappear in Cu_2(CuTCPP). The peak at around 1400 cm^{-1} in Cu_2(CuTCPP) is attributed the stretching mode of coordinated COO^- [20]

Fig. 2.6 **a** Scanning electron microscope (SEM) image of $Cu_2(CuTCPP)$ nanosheets. Transmission electron microscopy images of **b** $Cu_2(CuTCPP)$

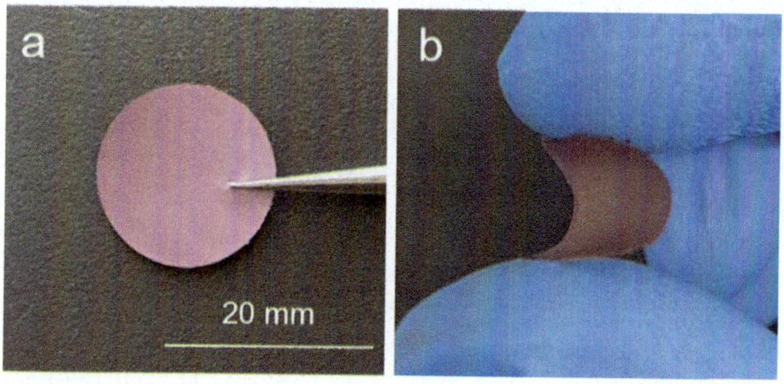

Fig. 2.7 **a** Photograph of $Cu_2(CuTCPP)$ free-standing membrane, flexibility of **b** $Cu_2(CuTCPP)$ free-standing membranes with a diameter of 17 mm

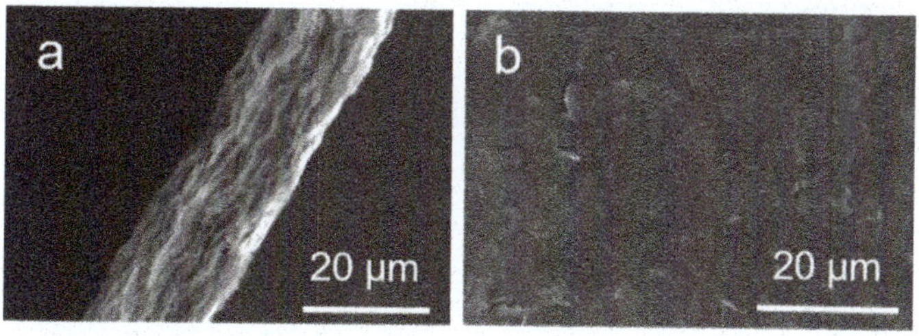

Fig. 2.8 **a** SEM image of the cross-section of the $Cu_2(CuTCPP)$ membrane, and **b** membrane surfaces

are well oriented in a parallel fashion (Fig. 2.12a), as evidenced by the (001) reflections in the out-of-plane XRD pattern (Fig. 2.13). Water vapor sorption isotherms of the membranes were measured to characterize their water adsorption behavior. As shown in Fig. 2.14, $Cu_2(CuTCPP)$ and Mg–Al–LDH(NO_3) membranes have high water vapor adsorption capacities of 105 and 484 $cm^3\ g^{-1}$, respectively, at $P/P_0 =$

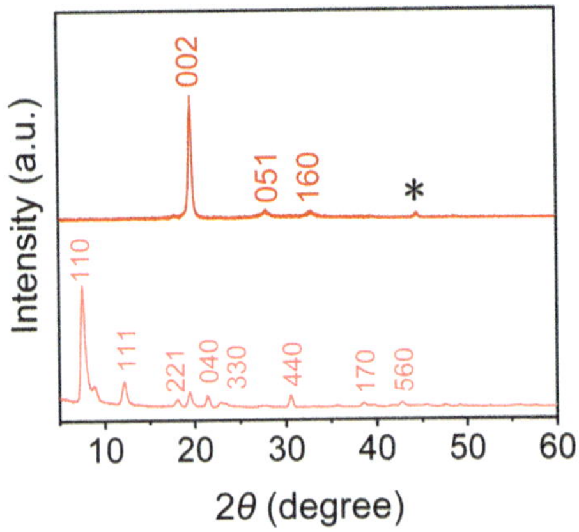

Fig. 2.9 Out-of-plane (red) and in-plane (pale red) X-ray diffraction (XRD) patterns of the Cu_2(CuTCPP) membrane (an asterisk denotes the peak from Si substrate)

0.92. There is no saturated adsorption platform, possibly owing to the irregular pore structures [21]. The sorption isotherm hysteresis observed in Mg–Al–LDH(NO_3) may reflect the possible chemisorption of water as described above [19]. As shown in Fig. 2.15a, the out-of-plane proton conductivity of the Cu_2(CuTCPP) membrane, which was measured by the a.c. impedance method, considerably increases with increasing humidity. This must be a reflection of the concentration of water in the pores, which can serve as hopping sites for protons originating from the uncoordinated carboxylic groups (Fig. 2.4) [22]. The Mg–Al–LDH(NO_3) membrane also showed humidity-dependent proton conductivity (Fig. 2.15b); however, the conductivity change is less pronounced than that of the Cu_2(CuTCPP) membrane. The presence of hydroxyl groups at higher concentrations than the proton species in Cu_2(CuTCPP) appears to be a part of the reason for humidity-independent conduction.

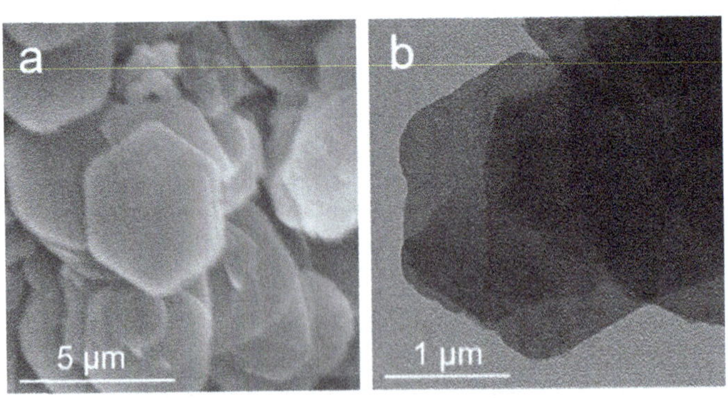

Fig. 2.10 **a** Scanning electron microscope (SEM) image of Mg–Al–LDH(NO_3) nanosheets. Transmission electron microscopy images of **b** Mg–Al–LDH(NO_3)

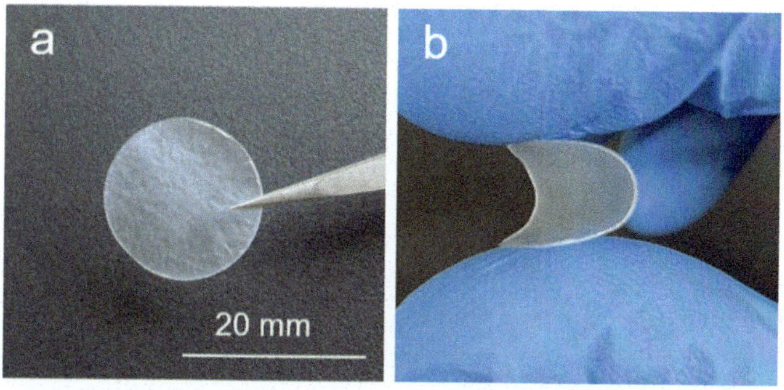

Fig. 2.11 **a** Photograph of Mg–Al–LDH(NO_3) free-standing membrane, flexibility of **b** Cu_2(CuTCPP) free-standing membranes with a diameter of 17 mm

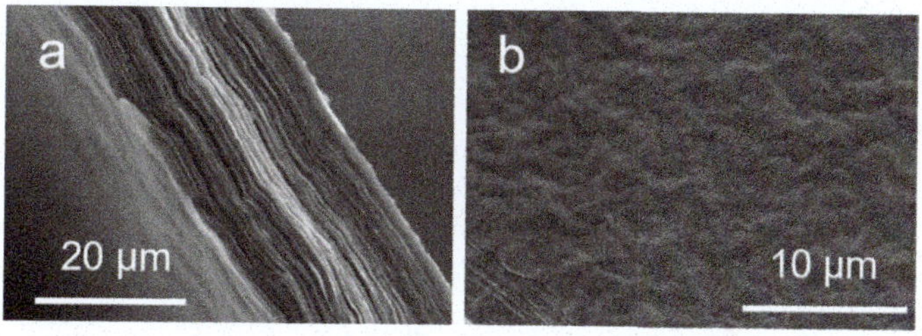

Fig. 2.12 **a** SEM image of the cross-section of the Mg–Al–LDH(NO_3) membrane, and **b** membrane surfaces

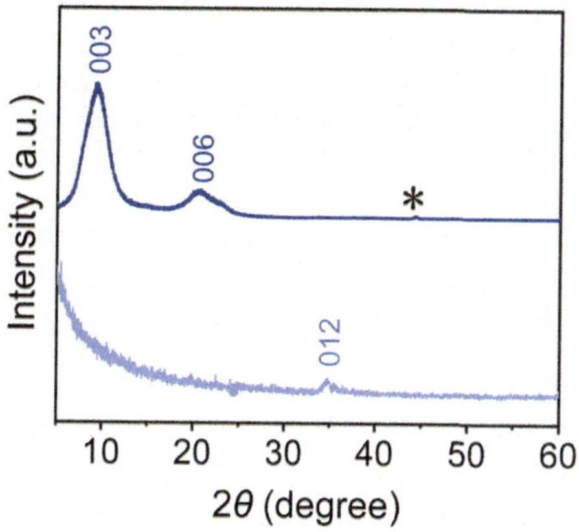

Fig. 2.13 Out-of-plane (red) and in-plane (pale red) X-ray diffraction (XRD) patterns of the Mg–Al–LDH(NO_3) membrane (an asterisk denotes the peak from Si substrate)

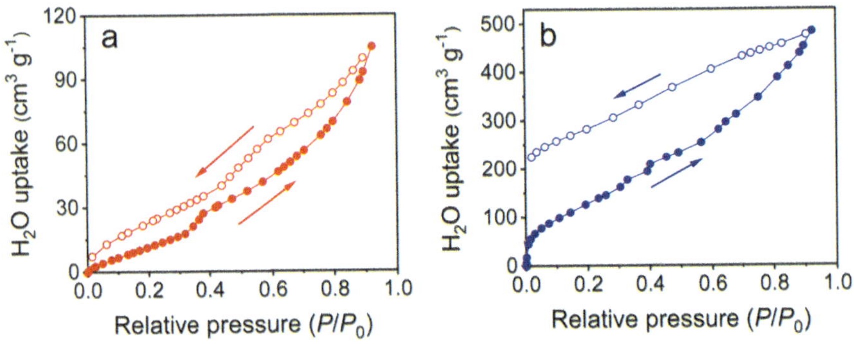

Fig. 2.14 Water vapor adsorption (closed circles) and desorption (open circles) isotherms of the pulverized **a** Cu_2(CuTCPP) and **b** Mg–Al–LDH(NO_3) membrane at 298 K

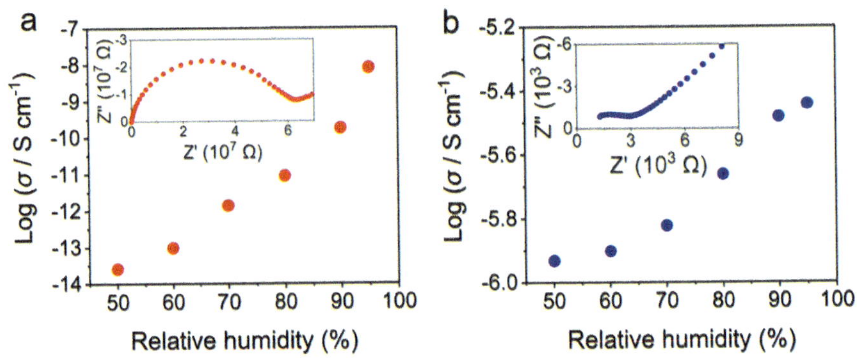

Fig. 2.15 Proton conductivity (σ) of Cu_2(CuTCPP) **a** and Mg–Al–LDH(NO_3) **b** membrane at 25 °C under various RH conditions (inset: Nyquist plot under 90% RH)

To investigate unidirectional ion conduction in the membranes, we performed d.c. resistivity measurements using Pd plates as electrodes at 25 °C and 90% RH. As shown in Fig. 2.16, a nearly linear and symmetric current–voltage (*I*–*V*) plot was observed when only one of the membranes, i.e., Cu_2(CuTCPP) or Mg–Al–LDH(NO_3), was sandwiched by the electrodes. Specifically, both membranes exhibited apparent relaxation behavior (Fig. 2.17) and significantly reduced the current under dry conditions (30% RH; Fig. 2.17). These results provide firm evidence that ions (i.e., protons and hydroxide ions) instead of electrons are responsible for the observed current in such membranes.

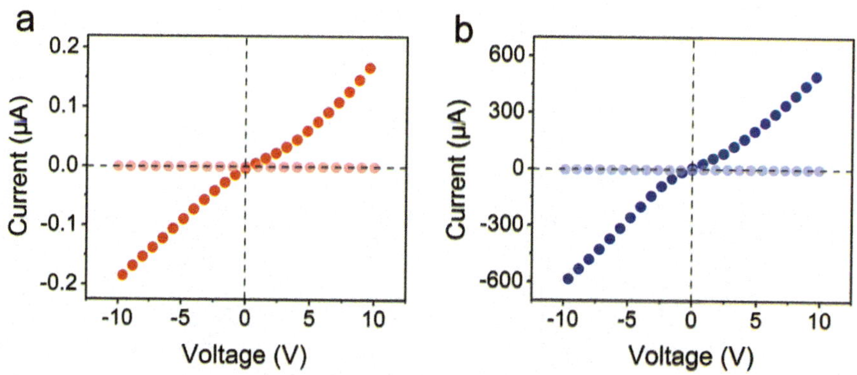

Fig. 2.16 *I–V* plots of single membranes of **a** $Cu_2(CuTCPP)$ and **b** Mg–Al–LDH(NO_3) at 25 °C under 30% RH (pale red/blue circles) and 90% RH (red/blue circles)

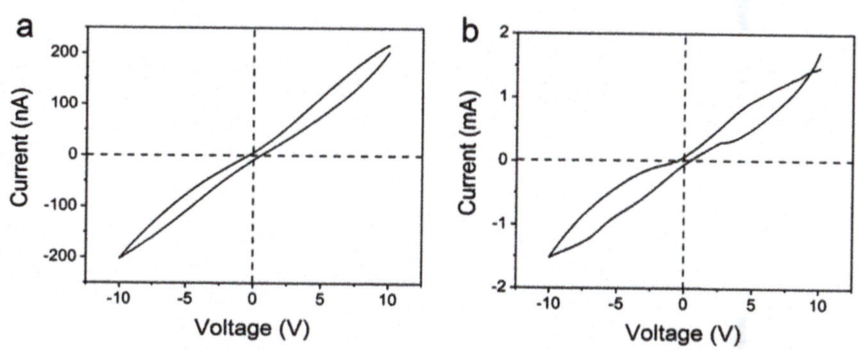

Fig. 2.17 Current–voltage cyclic curves of **a** $Cu_2(CuTCPP)$ and **b** Mg–Al–LDH(NO_3) membranes at 25 °C and 90% RH

The mechanical combination of the two membranes led to the drastic change in the *I–V* curve. As demonstrated in Fig. 2.18a, the heterogeneous membrane displayed a nonlinear and asymmetric *I–V* curve, where the forward bias current starts to increase at around 1.0 V. The highly rectified ion transport (623 μA at a forward bias voltage of +10 V and − 2.1 μA at a reverse bias voltage of − 10 V) clearly indicates unidirectional proton and hydroxide ion transport in the heterogeneous membrane. The proton rectification ratio (PRR), defined as the ratio of the forward/reverse bias current at ± 10 V, was estimated to be as high as 297. We also characterized unidirectional proton/hydroxide ion conduction by applying constant positive and negative d.c. voltages to the heterogeneous membrane. Figure 2.19b clearly demonstrates that the current becomes almost constant after applying a voltage of ± 10 V for approximately 10 s in both the forward and reverse directions. The mechanism of the continuous supply of protons and hydroxide ions will be discussed later. It should be noted that the PRR value estimated from the currents in the stable region (approximately 292) is in good agreement with that estimated based on the *I–V* curve.

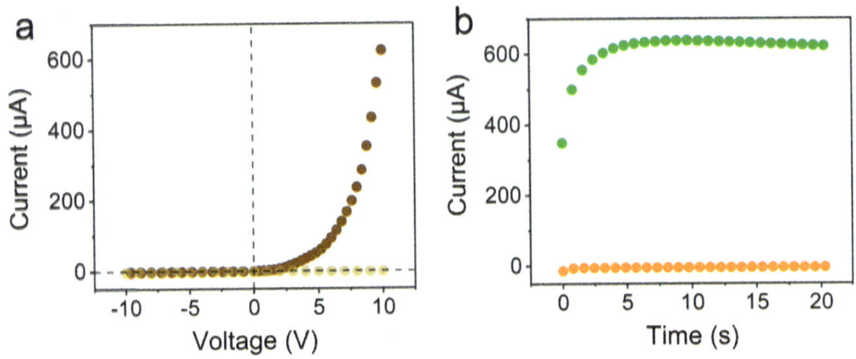

Fig. 2.18 **a** *I–V* plot of $Cu_2(CuTCPP)$/Mg–Al–LDH(NO_3) heterogeneous membrane at 25 °C under 30% RH (pale brown circles) and 90% RH (brown circles). **b** Time dependence of current of the heterogeneous membrane under applied voltages of + 10 V (forward bias; green circles) and − 10 V (reverse bias; orange circles)

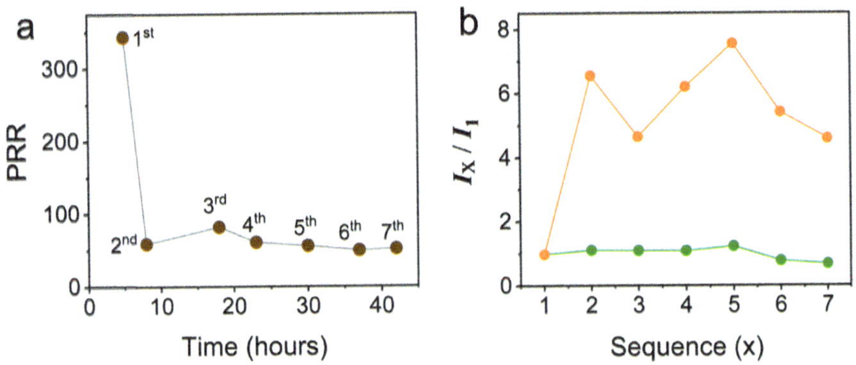

Fig. 2.19 **a** Sequence-dependent proton rectification ratio (PRR) of the heterogeneous membrane. **b** Relative forward (green circles) and reverse currents (orange circles) as a function of the number of scans (x) with respect to those in the first scan (I_x/I_1)

To assess the cyclability and reproducibility of the rectification performance, *I–V* measurements were conducted several times at intervals of several hours for seven different sets of heterogeneous membranes. A typical temporal variation of the PRR is shown in Fig. 2.21. The first scan yielded a very high PRR value (344), as shown in Fig. 2.19a; however, the value suddenly dropped to 59 in the second scan and remained fairly constant (50–82) in the following scan. Given that the membranes demonstrated a significant increase in the reverse bias current while maintaining the forward bias current in the second scan (Fig. 2.19b), the fall-off of the PRR possibly arises from the generation of water in the interfacial region between the $Cu_2(CuTCPP)$ and Mg–Al–LDH(NO_3) membranes during the first scan, owing to the association of migrated protons and hydroxide ions. Actually, we observed water streaks at the interface between the membranes after the experiments, which were invisible in the heterogeneous membrane placed at 25 °C under 90% RH for 2 days without applying a bias voltage (Fig. 2.20). The generation of water molecules consequently increases the reverse bias current because of the increased ion carrier concentration via water dissociation [3]. A similar behavior was observed in the other six heterogeneous membranes, as shown in Fig. 2.21 and Table 2.1, where the PRR values after the second scan lie in the range of 40–80. We note that five out of seven heterogeneous membranes presented a PRR value higher than 200 in the first scan, although the value varied considerably between the samples (35–922). Detailed see Figs. 2.22, 2.23, 2.24, 2.25, 2.26, 2.27, and 2.28.

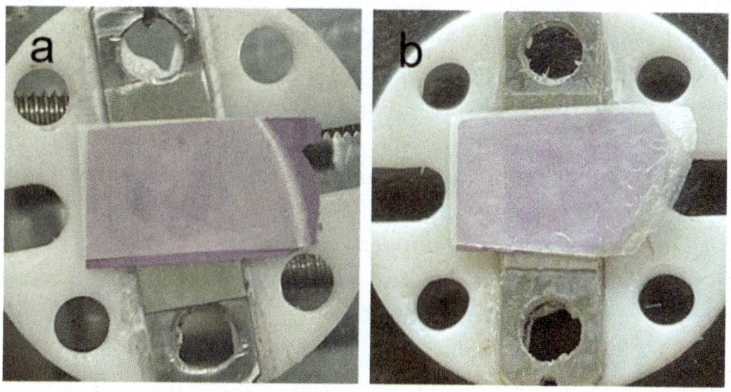

Fig. 2.20 Heterogeneous membrane **a** after a series of rectification experiments (25 °C and 90% RH for 2 days) and **b** after being placed under the same conditions without applying voltage (as a control experiment)

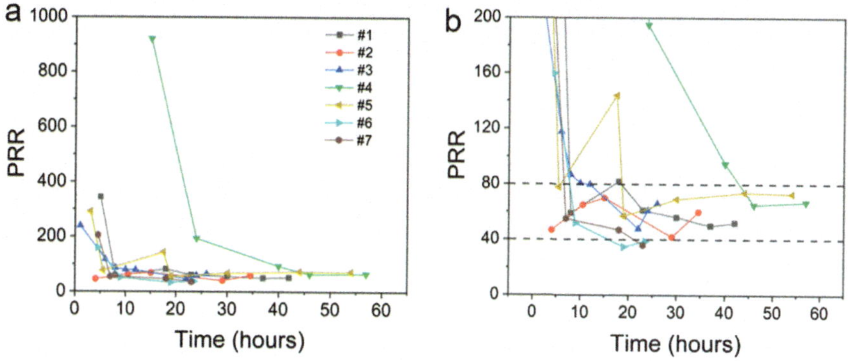

Fig. 2.21 **a** Temporal variation of proton rectification ratio (PRR) of seven kinds of heterogeneous membranes at 25 °C and 90% RH and **b** enlargement of Figure (**a**)

Table 2.1 Sequence-dependent PRR of seven kinds of heterogeneous membranes proton rectifiers

	Sequence							
	1st	2nd	3rd	4th	5th	6th	7th	8th
#1	344	59	82	61	56	50	52	
#2	47	65	70	42	60			
#3	240	118	87	81	80	48	61	66
#4	922	195	95	65	67			
#5	292	78	144	57	69	74	73	
#6	160	52	35	39				
#7	205	55	47	36				

The PXRD measurements revealed that the structure of each membrane was mostly intact during the rectification experiments (Fig. 2.29). As mentioned above, the operating principles of proton rectifiers are analogous to those of p–n diodes (Fig. 2.1). Because considerable amounts of protons and hydroxide ions are accommodated in the Cu_2(CuTCPP) and Mg–Al–LDH(NO_3) membranes, respectively, the resulting pH gradient promotes the diffusion of protons in the Cu_2(CuTCPP) membrane and hydroxide ions in the Mg–Al–LDH(NO_3) membrane toward the interfacial region, even under no applied voltage. The recombination of protons and hydroxide ions generates water, and the neutralized interfacial region (termed as the depletion layer) leads to current ceasing. Applying a forward bias voltage (i.e., Cu_2(CuTCPP) facing the anode) further promotes the ion diffusion of both protons and hydroxide ions toward the interfacial region, which is a factor dominating the observed current. The water electrolysis reactions at the boundary between the membrane and Pd electrode, i.e., water oxidation ($2H_2O \rightarrow O_2 + 4H^+ + 4e^-$) on the anode side and reduction on the cathode side ($2H_2O + O_2 + 4e^- \rightarrow 4OH^-$), presumably occur continuously to supply protons and hydroxide ions to the membranes (right panel of Fig. 2.30). In the opposing mode (i.e., Cu_2(CuTCPP) facing the cathode), the reverse bias increases the potential barrier for both protons and hydroxide ions, which suppresses ion diffusion driven by the concentration gradients. Eventually, protons and hydroxide ions are directed toward the cathode and anode sides, respectively. In the reverse bias mode, the dissociation of water accommodated in the interfacial region possibly generates protons and hydroxide ions (left panel of Fig. 2.30). The reverse bias current must be significantly low because of the low dissociation rate despite the generation of water in the interfacial region during the first scan, which appears to increase the reverse bias current.

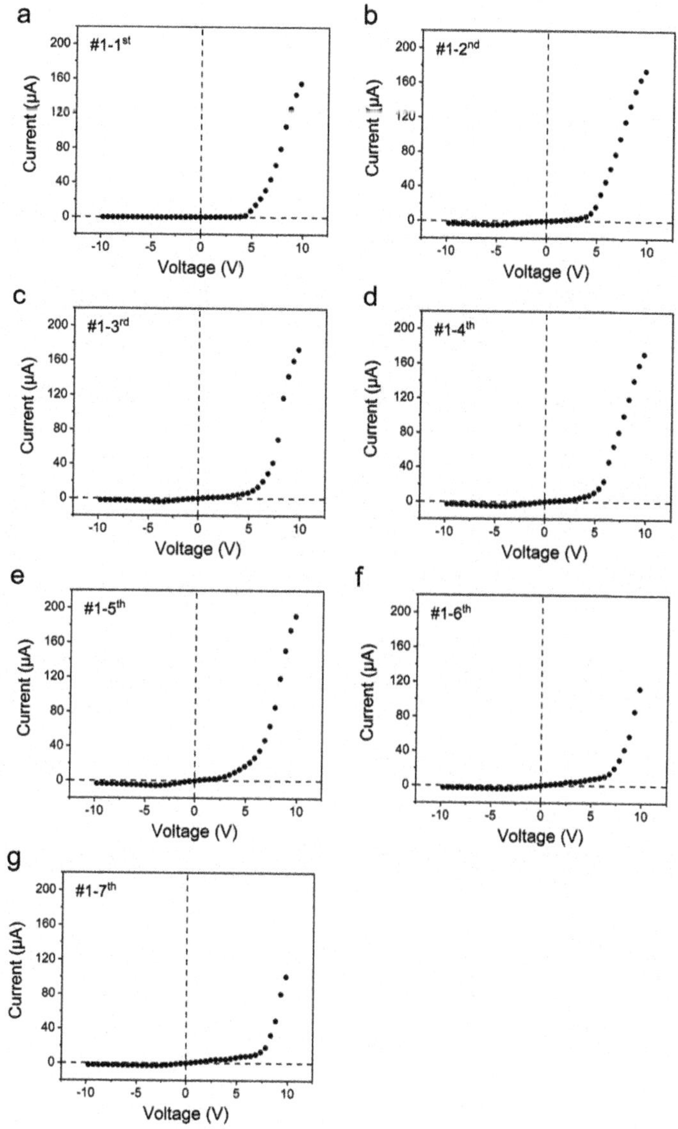

Fig. 2.22 *I–V* plots of sample #1 measured at different sequences

To date, a few all-solid-state H^{+}–OH^{-} rectifier devices have been fabricated using heterogeneous junctions of proton-conducting and hydroxide ion-conducting membranes. An early example is the work of Seno and Yamabe in 1964; acid- and base-washed ion-exchange resins were used after soaking in water (water weight: 56%). They demonstrated that a PRR value of approximately 100 was attained in a humidity-controlled chamber [9]. In 2013, Rolandi et al. reported the rectifying

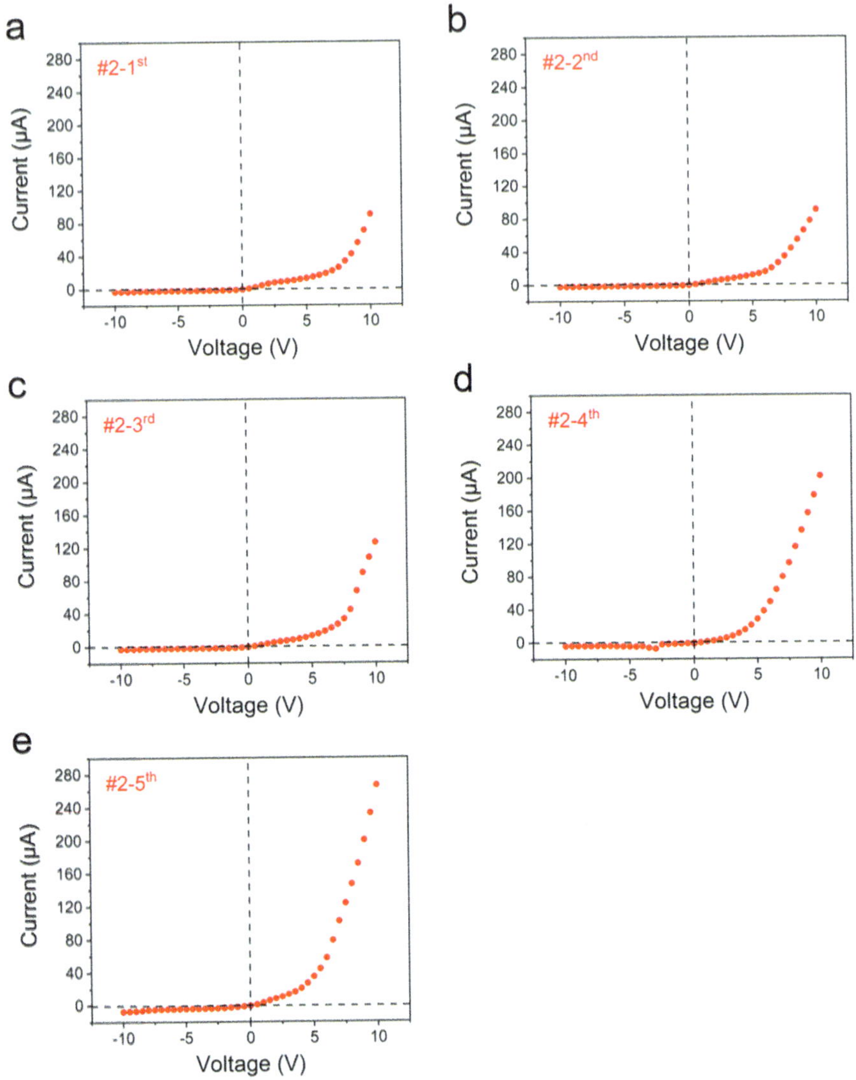

Fig. 2.23 *I–V* plots of sample #2 measured at different sequences

behavior of a heterogeneous membrane composed of maleic and proline chitosan films, which Table 2.2. Comparison of rectification results with the reported all-solid-state proton rectifiers presented a PRR value less than 10 under 75% RH and 5% H_2 gas atmosphere [9]. It should be noted that our all-solid-state H^+–OH^- rectifier device mainly demonstrates three advantages over those in the previous studies: (1) high

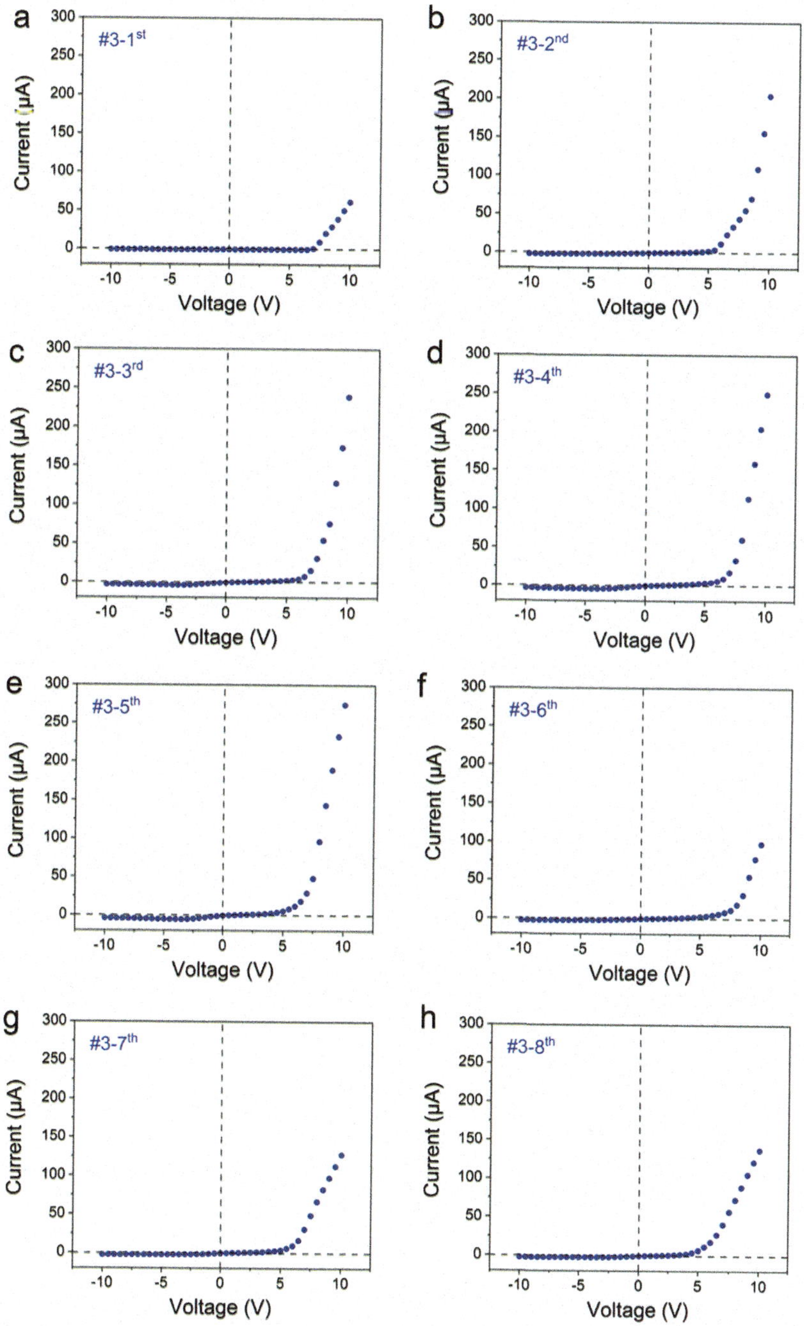

Fig. 2.24 *I–V* plots of sample #3 measured at different sequences

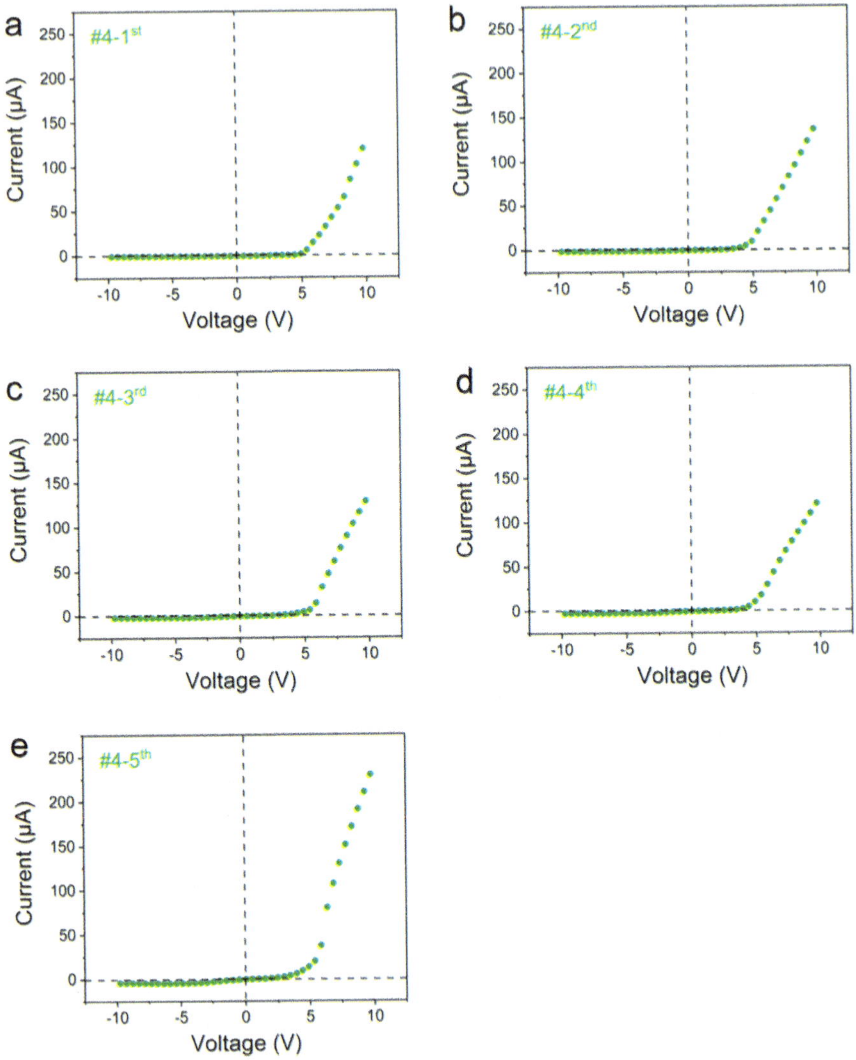

Fig. 2.25 *I–V* plots of sample #4 measured at different sequences

rectification ratio (PRR > 200), (2) high designability, and (3) well-defined structure of the device components. In addition, the facile operating conditions (i.e., 90% RH), without soaking in water prior to the performance or using flammable H_2 gas, render it particularly suitable for several practical applications. Notably, both membranes used in this study remained intact upon prolonged immersion in water (Fig. 2.31).

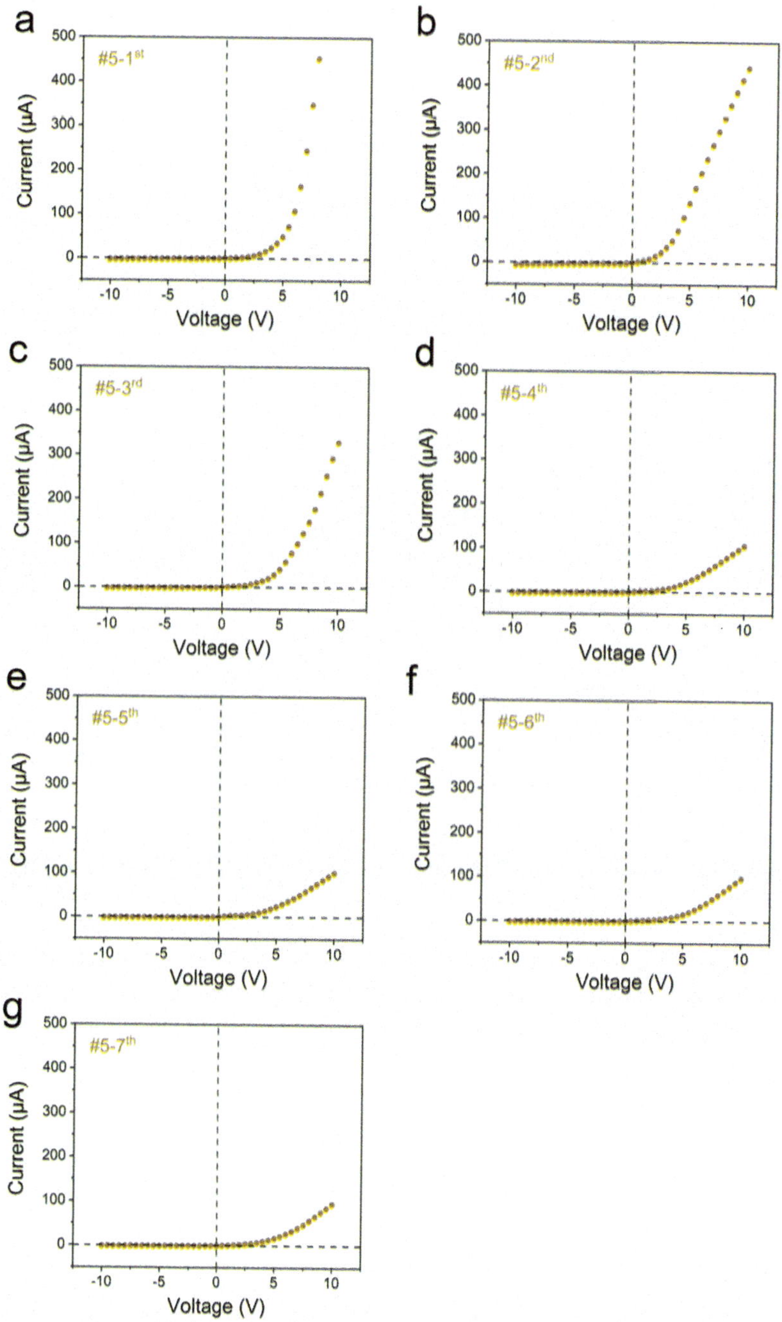

Fig. 2.26 *I–V* plots of sample #5 measured at different sequences

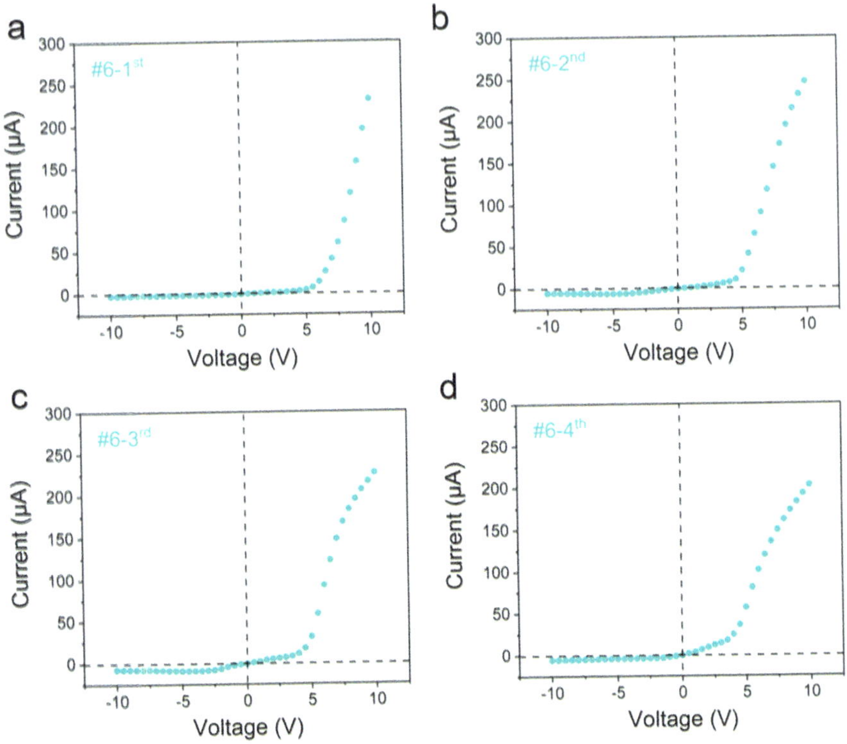

Fig. 2.27 *I–V* plots of sample #6 measured at different sequences

Chemical characteristics of PCPs, such as proton concentration (i.e., pK_a), proton mobility, and hydrophilicity/hydrophobicity, can be modified by selecting metal ions and organic linkers while maintaining the basic framework structure [13]. Given that their structures can be evaluated appropriately by diffraction techniques in contrast to the organic polymers used in earlier studies, the present system is evidently a promising platform for studying the structure–property relationship to develop all-solid-state H^+–OH^- rectifiers.

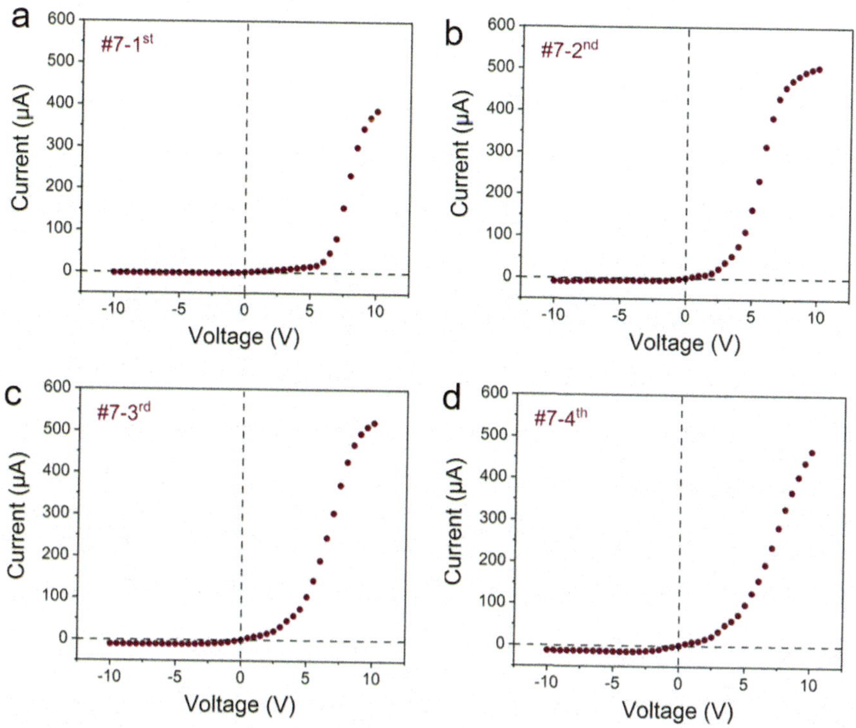

Fig. 2.28 *I–V* plots of sample #7 measured at different sequences

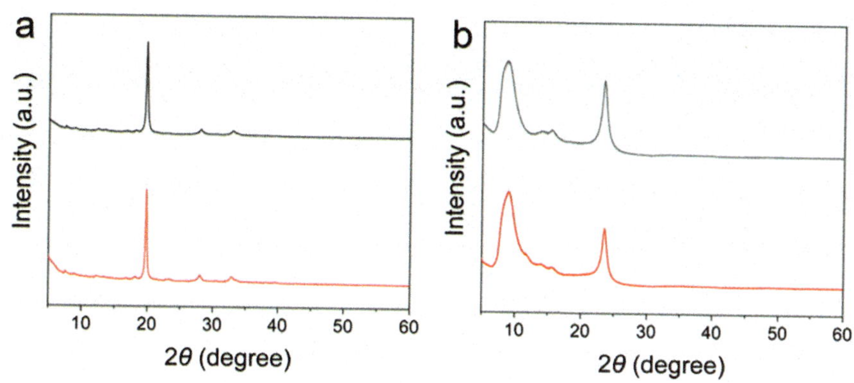

Fig. 2.29 PXRD patterns of **a** Cu_2(CuTCPP) and **b** Mg–Al–LDH(NO_3) membranes before (black) and after (red) rectification experiments

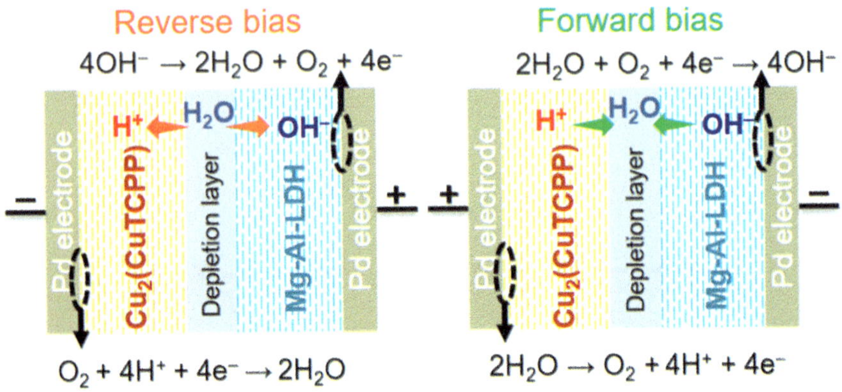

Fig. 2.30 Schematics of ion generation in the heterogeneous membrane under reverse and forward bias conditions

Table 2.2 Comparison of rectification results with the reported all-solid-state proton rectifiers

Proton conductor	Hydroxide ion conductor	Conditions	PRR	Ref.
Amberlite IR-120	Amberlite IRA-400	56% weight of water content	~100	[9a]
Dowex-50	Dowex-1	Conditioned in high RH for 12 h	~6	[9b]
Maleic chitosan	Proline chitosan	5% H_2, 75% RH	~6	[9d]
Cu_2(CuTCPP)	Mg–Al–LDH(NO_3)	90% RH	> 200	Our work

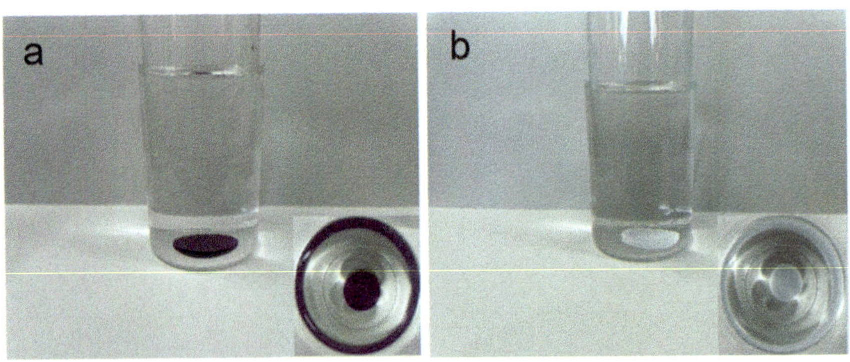

Fig. 2.31 Water stability tests for **a** Cu_2(CuTCPP) and **b** Mg–Al–LDH(NO_3) membranes, which remain intact in water after 40 days (insets are the top view images of the vial)

2.4 Conclusion

In conclusion, we fabricated an all-solid-state proton rectifier by the facile mechanical lamination of two free-standing membranes, namely, a proton-conducting PCP, Cu_2(CuTCPP), and hydroxide ion-conducting LDH, Mg–Al–LDH(NO_3). The *I*–*V* characteristics provide a rectification ratio as high as > 200 at 25 °C under 90% RH, which is the highest recorded value. In addition to the remarkable performance under electrolyte-free conditions, the present heterogeneous membrane has the potential to modify and control the water permeability, H^+/OH^- mobility, and pK_a in the pores, which noticeably affect the rectifying behavior, by selecting components such as metal ions and organic linkers. These results will open the way for the exploration of new and more advanced all-solid-state proton rectifiers through rational design. Studies along this line are in progress.

References

1. (a) H.C. Blair, S.L. Teitelbaum, R. Ghiselli, S. Gluck, Science **245,** 855–857 (1989); (b) G. Steinberg-Yfrach, J.L. Rigaud, E.N. Durantini, A.L. Moore, D. Gust, T.A. Moore, Nature **392,** 479–482 (1998); (c) D. Gadsby, Nat. Rev. Mol. Cell Biol. **10,** 344–352 (2009)
2. (a) D. Gust, T.A. Moore, A.L. Moore, Acc. Chem. Res. **34,** 40–48 (2001); (b) P.K. Ghosh, A.Y. Smirnov, F. Nori, J. Chem. Phys. **131,** 035102 (2009); (c) S. Berardi, S. Drouet, L. Francàs, C. Gimbert-Suriñach, M. Guttentag, C. Richmond, T. Stoll, A. Llobet, Chem. Soc. Rev. **43,** 7501–7519 (2014); (d) S. Chabi, K.M. Papadantonakis, N. S. Lewis, M.S. Freund, Energy Environ. Sci. **10,** 1320–1338 (2017)
3. (a) Y. Tanaka, Membr. Sci. Tech. **12,** 405–436 (2007); (b) R. Pärnamäe, S. Mareev, V. Nikonenko, S. Melnikov, N. Sheldeshov, V. Zabolotskii, H.V.M. Hamelers, M. Tedesco, J. Membr. Sci. **617,** 118538 (2021)
4. K. Jiao, J. Xuan, Q. Du, Z. Bao, B. Xie, B. Wang, Y. Zhao, L. Fan, H. Wang, Z. Hou, S. Huo, N.P. Brandon, Y. Yin, M.D. Guiver, Nature **595**, 361–369 (2021)
5. W. Guo, Y. Tian, L. Jiang, Acc. Chem. Res. **46**, 2834–2846 (2013)
6. B. Lovrecek, A. Despic, J.O.M. Bockris, J. Phys. Chem. **63**, 750–751 (1959)
7. (a) P.K. Giesbrecht, M.S. Freund, Chem. Mater. **32,** 8060–8090 (2020); (b) Z. Zhang, X.Y. Kong, K. Xiao, Q. Liu, G. Xie, P. Li, J. Ma, Y. Tian, L.P. Wen, L. Jiang, J. Am. Chem. Soc. **137,** 14765–14772 (2015)
8. (a) Z. Siwy, A. Fuliński, Am. J. Phys. **72,** 567–574 (2004); (b) L.J. Cheng, L.J. Guo, Chem. Soc. Rev. **39,** 923–938 (2010)
9. (a) M. Seno, T. Yamabe, Bull. Chem. Soc. Jpn. **37,** 668–671 (1964); (b) J.J. Langer, Appl. Phys. A **34,** 195–198 (1984); (c) J.J. Langer, M. Martyński, Adv. Opt. Mater. **9,** 15–18 (1999); (d) Y. Deng, E. Josberger, J. Jin, A.F. Roudsari, B.A. Helms, C. Zhong, M.P. Anantram, M. Rolandi, Sci. Rep. **3,** 2481 (2013)
10. (a) A. Schoedel, M. Li, D. Li, M. O'Keeffe, O.M. Yaghi, Chem. Rev. **116,** 12466–12535 (2016); (b) L. Feng, K.Y. Wang, G.S. Day, H.C. Zhou, Chem. Soc. Rev. **48,** 4823–4853 (2019); (c) S. Kitagawa, R. Matsuda, Coord. Chem. Rev. **251,** 2490–2509 (2007)
11. (a) K.D. Kreuer, Chem. Mater. **8,** 610–641 (1996); (b) *Proton Conductors: Solids, Membranes and Gels–Materials and Devices*, ed. by P. Colomban (Cambridge University Press, Cambridge, 1992)
12. (a) D. Zhao, D.J. Timmons, D. Yuan, H.C. Zhou, Acc. Chem. Res. **44,** 123–133 (2011); (b) M.Y. Masoomi, A. Morsali, A. Dhakshinamoorthy, H. Garcia, Angew. Chem. Int. Ed. **58,**

15188–15205 (2019), Angew. Chem. **131,** 15330–15347 (2019); (c) A. Kirchon, L. Feng, H.F. Drake, E.A. Joseph, H.C. Zhou, Chem. Soc. Rev. **47,** 8611–8638 (2018); (d) A.E. Baumann, D.A. Burns, B. Liu, V.S. Thoi, Commun. Chem. **2,** 86 (2019); (e) H.D. Lawson, S.P. Walton, C. Chan, ACS Appl. Mater. Interfaces **13,** 7004–7020 (2021); (f) W. Lu, Z. Wei, Z.Y. Gu, T.F. Liu, J. Park, J. Park, J. Tian, M. Zhang, Q. Zhang, T. Gentle III, M. Bosch, H.C. Zhou, Chem. Soc. Rev. **43,** 5561–5593 (2014); (g) R. Freund, O. Zaremba, G. Arnauts, R. Ameloot, G. Skorupskii, M. Dincă, A. Bavykina, J. Gascon, A. Ejsmont, J. Goscianska, M. Kalmutzki, U. Lächelt, E. Ploetz, C.S. Diercks, S. Wuttke, Angew. Chem. Int. Ed. **60,** 23975–24001 (2021), Angew. Chem. **133,** 24174–24202 (2021); (h) N. Stock, S. Biswas, Chem. Rev. **112,** 933–969 (2012); (i) J. Li, P.M. Bhatt, J. Li, M. Eddaoudi, Y. Liu, Adv. Mater. **32,** 2002563 (2020)

13. D.W. Lim, H. Kitagawa, Chem. Rev. **120**, 8416–8467 (2020)
14. Makiura, S. Motoyama, Y. Umemura, H. Yamanaka, O. Sakata, H. Kitagawa, Nat. Mater. **9,** 565–571 (2010)
15. (a) G. Xu, T. Yamada, K. Otsubo, S. Sakaida, H. Kitagawa, J. Am. Chem. Soc.**134,** 16524–16527 (2012); (b) G. Xu, K. Otsubo, T. Yamada, S. Sakaida, H. Kitagawa, J. Am. Chem. Soc. **135,** 7438–7441 (2013); (c) M. Tian, F. Pei, M. Yao, Z. Fu, L. Lin, G. Wu, G. Xu, H. Kitagawa, X. Fang, Energy Storage Mater. **21,** 14–21 (2019); (d) G.D. Wu, H.L. Zhou, Z.H. Fu, W.H. Li, J.W. Xiu, M.S. Yao, Q.H. Li, G. Xu, Angew. Chem. Int. Ed. **60,** 9931–9935, Angew. Chem. **133,** 10019–10023 (2021)
16. (a) X. Huang, Z. Zeng, H. Zhang, Chem. Soc. Rev. **42,** 1934–1946 (2013); (b) H. Zhang, Z.H. Fu, D. Legut, T.C. Germann, R.F. Zhang, RSC Adv. **7,** 55912–55919 (2017)
17. K.H. Goh, T.T. Lim, Z. Dong, Water Res. **42**, 1343–1368 (2008)
18. (a) P. Sun, R. Ma, X. Bai, K. Wang, H. Zhu, T. Sasaki, Sci. Adv. **3,** e1602629 (2017); (b) Y. Furukawa, K. Tadanaga, A. Hayashi, M. Tatsumisago, Solid State Ionics **192,** 185–187 (2011); (c) L. Li, R. Ma, Y. Ebina, N. Iyi, T. Sasaki, Chem. Mater. **17,** 4386–4391 (2005)
19. P. Sun, F. Chen, W. Zhou, X. Liu, R. Ma, T. Sasaki, Mater. Horiz. **6**, 2087–2093 (2019)
20. *Infrared and Raman Spectra of Inorganic and Coordination Compounds, Part B: Applications in Coordination, Organometallic, and Bioinorganic Chemistry*, ed. by K. Nakamoto (Wiley, Hoboken, 2009)
21. D.G. Evans, X. Duan, Chem. Commun. **5**, 485–496 (2006)
22. Y.N. Zhou, L.L. Liu, Q.W. Liu, X.X. Liu, M.Z. Feng, L. Wang, Z.G. Sun, Y.Y. Zhu, X. Zhang, C.Q. Jiao, Inorg. Chem. **60**, 17303–17314 (2021)

Chapter 3
Robust Proton Conduction Against Mechanical Stress in Flexible Free-Standing Membrane Composed of Two-Dimensional Coordination Polymer

Abstract Introduction of mechanical flexibility into proton-conducting coordination polymers (CPs) is in high demand for future protonic applications such as fuel cells and hydrogen sensors. Whereas such mechanical properties have been primarily investigated in one-dimensional (1D) CPs, in this study we successfully fabricated highly flexible free-standing-type CP membranes with a high surface-to-volume ratio, which is beneficial for the increased performance in the foregoing applications. We employed a layered CP, Cu_2(NiTCPP) (H_4(H_2TCPP); 5,10,15,20-tetrakis(4-carboxyphenyl)porphyrin), in which a two-dimensional (2D) square grid sheet composed of tetradentate nickel porphyrins and paddlewheel-type copper dimers is connected to each other by weak van der Waals forces. Mechanical flexibility was evaluated by bending and tensile tests; flexural and Young's moduli of the membrane are significantly higher than that of conventional Nafion membranes. Electrochemical impedance spectroscopy revealed that the in-plane proton conductivity of the membrane was almost kept even by applying bending stress. Because the X-ray diffraction analysis indicates that proton-conducting pathway through the hydrogen-bonding network is kept intact during the bending operation, our present study provides a promising strategy for fabricating new and more advanced 2D CPs without using any substrates and addition polymers for protonic devices.

Keywords Coordination polymers · Free-standing membrane · Mechanical flexibility · Protonic device

3.1 Introduction

Porous coordination polymers (PCPs) are crystalline materials, which are essentially formed by self-assembling metal ions/clusters with organic linkers via coordination bonds [1]. Because the judicious selection of the components can control the surface potential (i.e., hydrophilicity/hydrophobicity and Lewis acidity/basicity) and geometrical structure of the pore spaces, PCPs have provided a versatile platform for

J. Lu, *Novel Proton Conducting in Free-Standing Coordination Polymer Membranes*, Springer Theses, https://doi.org/10.1007/978-981-97-7913-0_3

proton conduction when Lewis-acidic molecules (e.g., phosphoric acid) or Lewis-basic molecules (e.g., water) are accommodated in the pores [2]. Therefore, PCPs have a great potential for protonic applications such as proton exchange membrane fuel cells [3], proton field-effect transistors [4], hydrogen sensors [5], and proton separation [6].

To date, most of the researches on the proton-conducting CPs have been conducted exclusively in their crystalline or pelletized form [7]. It is known that the introduction of flexibility, namely, excellent mechanical stability against external force such as compressive, tensile, and flexural forces into the CPs not only increases the feasibility of the foregoing applications but also expands the range of their practical use [8]. However, the framework structure formed by directional covalent or coordination bonds is fraught with drawbacks of the difficulty in the introduction of mechanical flexibility [9]. To overcome the drawback, researchers have utilized low-dimensional CPs with open-framework structure, in which each chain or sheet unit is connected by weak van der Waals interactions. Whereas the flexural properties of needle-shaped crystalline CPs with one-dimensional (1D) coordination structure have been extensively investigated [10], mechanically flexible crystalline CPs with two-dimensional (2D) coordination structure have been almost unexplored despite their high surface-to-volume ratio is potentially beneficial for various applications mentioned above. Although there are several flexible polymer composites embedded with CPs [11], it is known that the chemical/physical properties of the composites depend largely on the kinds of the polymer and the loading amount of CPs [12]. Therefore, the realization of a flexible membrane composed solely of CPs is in high demand for not only the rational design of mechanically flexible proton-conducting CPs but also their future protonic applications. We note that a key issue to push forward the applications is to maintain the structural and proton-conducting properties during the repetitive sequences between flat and bending states; however, there is no report on how the application of the external force affects the proton conductivity of the 2D CPs.

We have focused on a proton-conducting 2D CP composed of tetradendate metal porphyrins and paddlewheel-type metal dimers connected by coordination bonds [13]. Highly flexible behavior, which must be a reflection of the layered structure formed by interlayer van der Waals interactions, Cu_2(CuTCPP) (H_4(H_2TCPP); 5,10,15,20-tetrakis(4-carboxyphenyl)porphyrin) membrane, was first found by Xu et al. [13c], whereas our group recently realized an all-solid proton-rectifying device with exceptionally high performance (rectification ratio > 200) using the Cu_2(CuTCPP) membrane [13d]. In this study, we investigated the proton-conducting behavior of a high-orientated free-standing membrane of the nickel substituent, Cu_2(NiTCPP), under applied bending stress. The excellent mechanical properties, which are visually manifested in the low limited bending radius, were quantitatively evaluated by the bending and tensile stress–strain tests. Our present study is the first to demonstrate that the proton conductivity of the membrane composed solely of CPs remains almost constant during the repetitive sequences between flat and bending states, which would be an important step toward future exploration of mechanically flexible 2D CPs.

3.2 Experimental

3.2.1 Materials

Copper(II) nitrate trihydrate (99%), nickel(II) acetate tetrahydrate (99%), and *N*,*N*-dimethylformamide (DMF, 99.5%) were used as received from Fujifilm Wako Pure Chemical Industries, Ltd. Tetrakis(4-carboxyphenyl)porphyrin ($H_4(H_2TCPP)$, 97%) was purchased from Tokyo Chemical Industry Co., Ltd., and used without further purification. Water was deionized before use.

3.2.2 Characterizations

Powder X-ray diffraction (PXRD) measurements were performed using a Bruker D8 ADVANCE with Cu Kα radiation ($\lambda = 0.15418$ nm) in the 2θ range from 5° to 60° with a 0.02° step and a step time of 1 s. In-plane and out-of-plane grazing-incidence XRD patterns were collected using a Rigaku Smart-Lab X-ray diffractometer with Cu Kα radiation ($\lambda = 0.15418$ nm) in the 2θ range from 5 to 50° at a scanning rate of 0.5° min^{-1}. Fourier-transform infrared (FTIR) spectra were obtained using a PerkinElmer Spectrum 100 spectrometer in attenuated total reflection (ATR) mode. Thermal stability was characterized by thermogravimetric (TG) measurements using a Bruker TG-DTA2000SA instrument at a heating rate of 5 °C min^{-1} under flowing N_2. The sorption experiments for N_2 (77 K) were performed using a BELSORP-max apparatus (MicrotracBEL). Ultraviolet–visible (UV–Vis) absorption spectra were measured using a JASCO V-570 UV–Vis-NIR spectrometer in a NaOH solution. Scanning electron microscope (SEM) images of the nanosheets and membranes were obtained using a Hitachi SU-1510 instrument. Transmission electron microscope (TEM) images of the nanosheets were obtained using a Hitachi HT7700 instrument. The strain–stress and flexural modulus experiments were performed by a Autograph AG-Xplus instrument. The impedance spectra of the membranes were measured using an Impedance/Gain-Phase Analyzer SI 1260 at an applied a.c. voltage of 100 mV. Gold paste (Tokuriki, 8560-1A) was painted on opposite sides of the membrane.

3.2.3 Preparation of H_4(NiTCPP) Ligand

According to the previous report [14], H_4(NiTCPP) ligand was prepared by reacting $H_2(H_4TCPP)$ (0.33 mmol) with $Ni(OAc)_2 \cdot 4H_2O$ (1.82 mmol) in DMF (70 mL) at reflux for 2 h. DMF was removed by distillation, and the crude H_4(NiTCPP) was precipitated by adding water, washed with water, and then collected by centrifugation. The product was dissolved in 100 mL of 0.1 M NaOH solution and reprecipitated by adding 20 mL of 1 M HCl solution. The precipitate was dissolved in ethanol and recrystallized by solvent evaporation (purple powder, 214.2 mg, yield: 76.7%). ^{1}H NMR (600 MHz, $CDCl_3$) δ 13.28 (s, 4H), 8.76 (s, 8H), 8.30 (d, 8H), 8.14 (d, 8H) ppm.

3.2.4 Preparation of Cu_2(NiTCPP) Nanosheets

Cu_2(NiTCPP) nanosheets were synthesized according to the previous procedure [13]. Typically, 8.5 mg (0.01 mmol) of H_4(NiTCPP) was dissolved into 13 mL of DMF, mixed with 5 mL of 1 mg mL^{-1} DMF solution of $Cu(NO_3)_2{\cdot}3H_2O$ (0.02 mmol), and ultrasonicated for 10 min. The suspension was aged at 35 °C for 10 h, heated at 85 °C for 24 h, and then cooled to room temperature within 12 h. Finally, the gel product was collected by centrifugation and washed with ethanol more than five times until the supernatant became colorless. The purple polycrystalline product was then dried at 50 °C for 12 h (8.0 mg, yield: 81.8%).

3.2.5 Preparation of Cu_2(NiTCPP) Free-Standing Membrane

Cu_2(NiTCPP) free-standing membrane was fabricated according to the previous procedure [13c, d]. Typically, 20 mg of polycrystalline Cu_2(NiTCPP) was added to 200 mL of ethanol in a 300 mL flask, which was tightly sealed after purging with N_2 gas. Ultrasonic treatment for 20 min allowed the exfoliation of the nanosheets to disperse well in ethanol. A Cu_2(NiTCPP) free-standing membrane was prepared by filtering the ethanolic suspension. Briefly, 20 mL of the ethanolic Cu_2(NiTCPP) suspension (0.2 mg mL^{-1}) was vacuum-filtered through a polyvinylidene difluoride (PVDF) membrane filter (pore diameter: ~0.22 μm; porosity: ~80%; diameter: 25 mm) and dried at 50 °C for 4 h. The Cu_2(NiTCPP) membrane with a thickness of approximately 8 μm could be easily detached from the filter to allow free-standing operation. Thickness of the membrane can be controlled by adjusting the loaded amount of the suspension for vacuum filtration process.

3.3 Results and Discussion

Cu_2(NiTCPP) nanosheets were prepared by stepwise synthesis. H_4(NiTCPP) was prepared by reacting *meso*-tetrakis(4-carboxyphenyl)porphyrin ($H_4(H_2TCPP)$) with $Ni(OAc)_2{\cdot}4H_2O$ according to the literatures (see Experimental part for details) [14]. The successful synthesis of H_4(NiTCPP) was verified by ultraviolet–visible (UV–vis) absorption (Fig. 3.2) and Fourier-transform infrared (FTIR) (Fig. 3.1) spectra in addition to ^{1}H nuclear magnetic resonance (NMR) spectra (Fig. 3.3). Purple crystalline Cu_2(NiTCPP) nanosheets were obtained by a solvothermal reaction in *N*,*N*-dimethylformamide (DMF) as reported previously (see Experimental part for details) [13c]. The powder X-ray diffraction (XRD) pattern of pulverized Cu_2(NiTCPP) nanosheets is consistent with the simulated pattern of the homobimetallic 2D CP, Cu_2(CuTCPP) (Fig. 3.4), with a layered structure [13]. As shown in Fig. 3.5, each Cu_2(NiTCPP) sheet has a square grid structure in the *ab* plane, which is composed of tetradentate nickel(II) porphyrin ($NiTCPP^{4-}$) and paddlewheel-type copper(II) dimers. Along the *c* axis, the reticulate layers are assembled mainly through van der Waals interactions in an AB stacking mode, where the B layer is shifted by approximately *a*/4 relative to the A layer. FTIR spectra showed the disappearance of the N–H bending mode (966 cm^{-1} in $H_4(H_2TCPP)$) in H_4(NiTCPP), indicating the metallization of the porphyrin units (Fig. 3.6). The disappearance of the C=O stretching mode in COOH groups (1676 cm^{-1} in H_4(NiTCPP)) in Cu_2(NiTCPP) affirmed that the four carboxyl groups of NiTCPP units successfully coordinate to the adjacent copper ions to form copper paddlewheel units [15]. X-ray photoelectron spectroscopy (XPS) detected the peaks characteristic of Cu and Ni atoms in Cu_2(NiTCPP) (Fig. 3.7); the peaks located at approximately 955 and 935 eV being assigned to Cu $2p_{1/2}$ and Cu $2p_{3/2}$, respectively [16], whereas those at approximately 873 and 856 eV being assigned to Ni $2p_{1/2}$ and Ni $2p_{3/2}$, respectively [17]. The Cu/Ni molar ratio estimated from the X-ray fluorescence (XRF) analysis (1.9) is consistent with that of the expected composition. The transmission electron microscope (TEM) image of Cu_2(NiTCPP), which was deposited onto a carbon-coated copper substrate by dropping the ethanol suspension, exhibited a nanosheet-like morphology (Fig. 3.8).

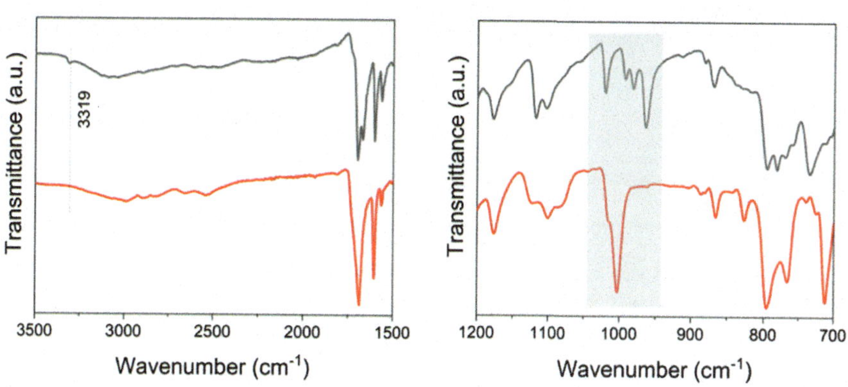

Fig. 3.1 FTIR spectra of $H_4(H_2TCPP)$ (black line) and H_4(NiTCPP) (red line) in the frequency regions of 1500–3500 cm^{-1} (top) and 700–1200 cm^{-1} (bottom). An N–H stretching band (3319 cm^{-1}), which is found in $H_4(H_2TCPP)$, disappears in Cu_2(NiTCPP). The number of peaks at around 1000 cm^{-1} decreases from four in $H_4(H_2TCPP)$ (N–H bending vibration) to one in H_4(NiTCPP) (N–Ni stretching vibration), which demonstrates that the porphyrin center in H_4(NiTCPP) has been filled with a nickel(II) ion

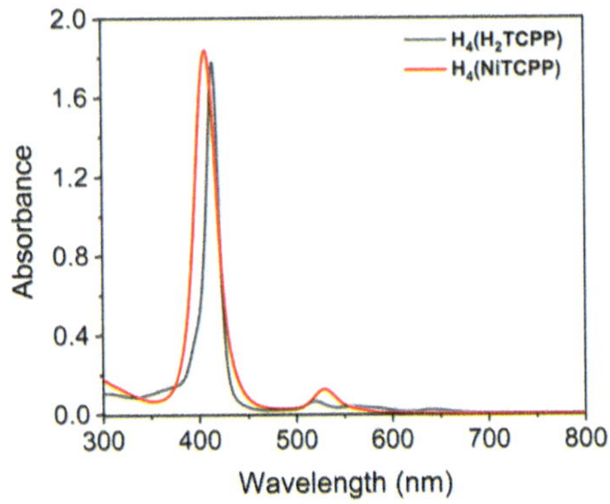

Fig. 3.2 UV–vis spectra of $H_4(H_2TCPP)$ (black line) and $H_4(NiTCPP)$ (red line) in a NaOH solution. The number of Q band peaks decreases from three in $H_4(H_2TCPP)$ to one in $H_4(NiTCPP)$, which demonstrates that the porphyrin center in $H_4(NiTCPP)$ is filled with a nickel(II) ion

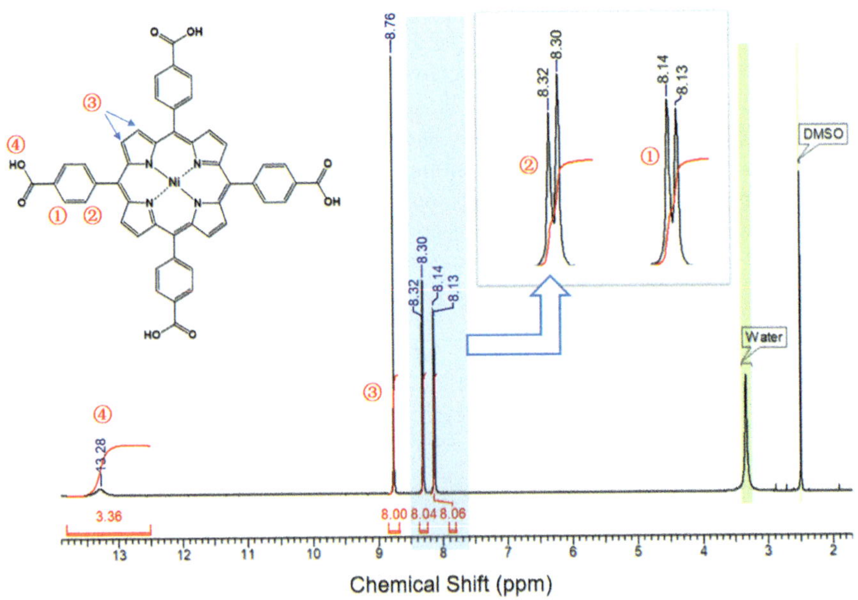

Fig. 3.3 ^1H NMR spectrum of $H_4(NiTCPP)$ in DMSO-d^6

Because the $Cu_2(NiTCPP)$ nanosheets were well dispersed in ethanol, the ethanol suspension was filtered through a polyvinylidene difluoride (PVDF) membrane filter to yield a highly oriented membrane. It is noteworthy that the membrane with high flexibility could be readily peeled from the filter after drying at 50 °C. Figure 3.9

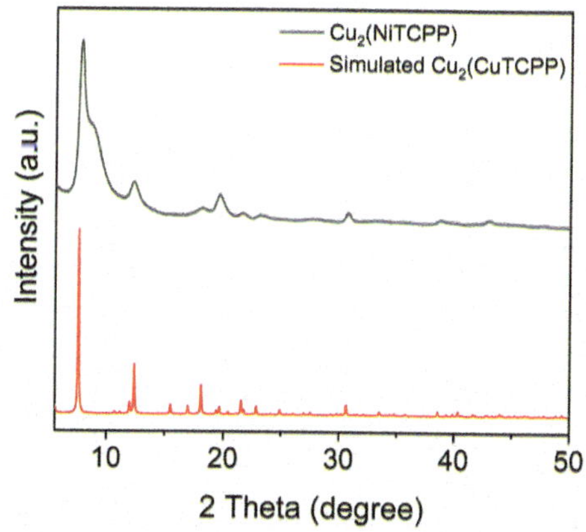

Fig. 3.4 PXRD pattern of pulverized Cu_2(NiTCPP) nanosheets (black) together with the simulated pattern of Cu_2(CuTCPP) (red) [3]

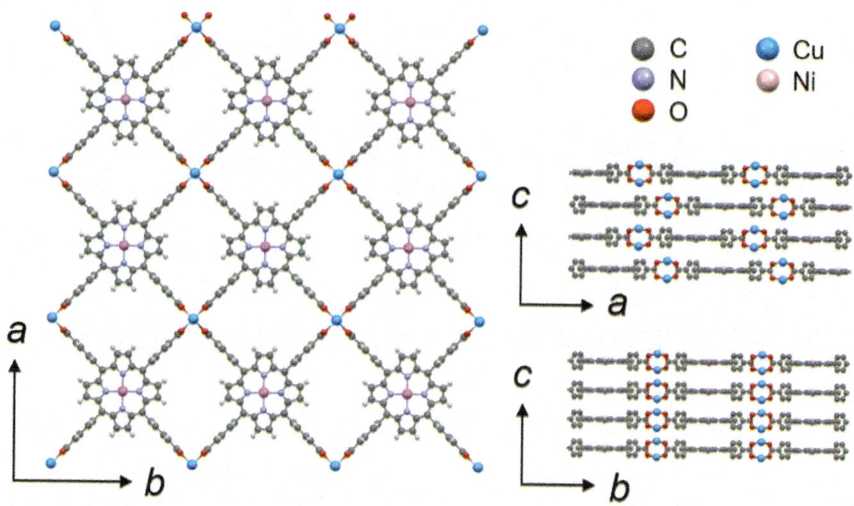

Fig. 3.5 Structure of Cu_2(NiTCPP)

shows a photograph of the free-standing membrane with a diameter of 17 mm and a thickness of approximately 8 μm. The thickness can be controlled in the range of less than ten to hundreds of micrometers by adjusting the loaded amounts of the suspension. The cross-sectional scanning electron microscope (SEM) image (Fig. 3.10) showed that the membrane is formed by stacking the nanosheets, which is consistent with the SEM image of the membrane surface. Figure 3.11 shows the out-of-plane and in-plane grazing-incidence XRD patterns of the membrane. It is obvious that the out-of-plane XRD pattern is dominated by an interlayer (002) reflection, indicating the preferred orientation of the membrane along the crystallographic *c* axis.

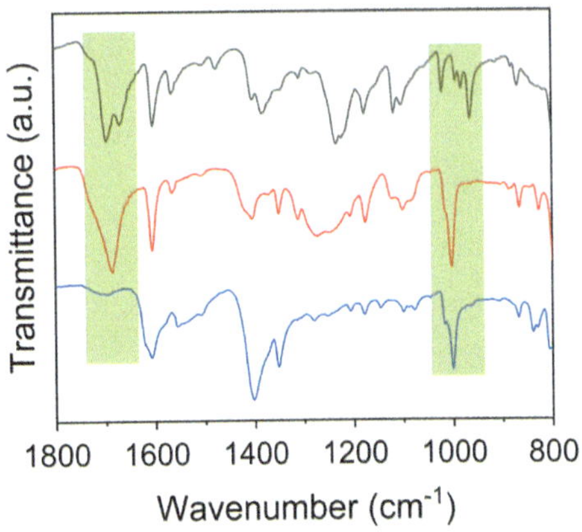

Fig. 3.6 Infrared spectra of $H_4(H_2TCPP)$ (black), $H_4(NiTCPP)$ (red), and $Cu_2(NiTCPP)$ (blue)

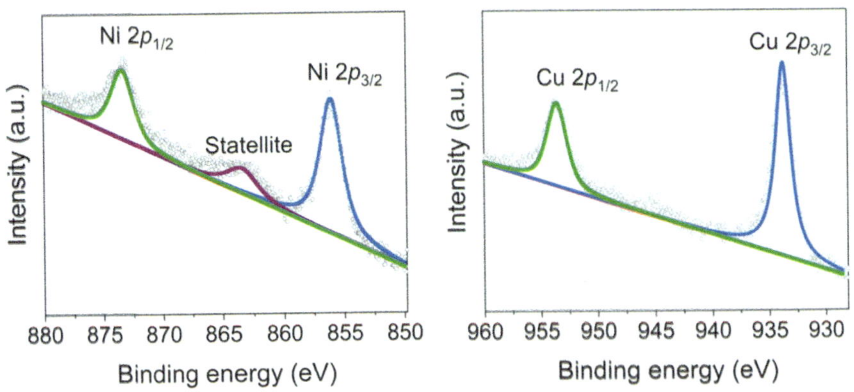

Fig. 3.7 X-ray photoelectron spectra of Ni 2*p* and Cu 2*p* regions in $Cu_2(NiTCPP)$ nanosheets, where the blue, green, and purple lines are fitted to the measured data (open circles)

Fig. 3.8 Transmission electron microscope image of $Cu_2(NiTCPP)$ nanosheets

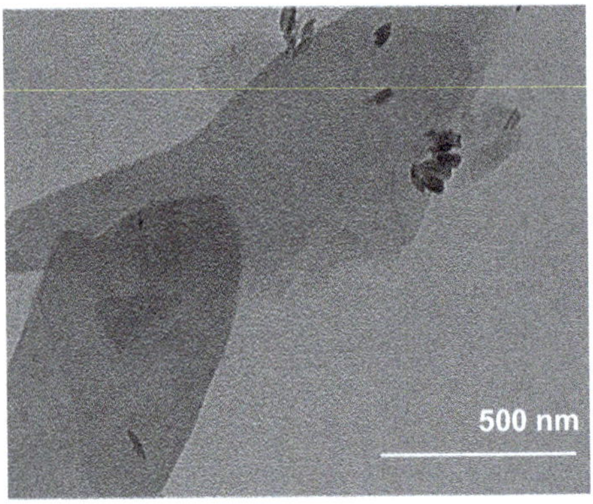

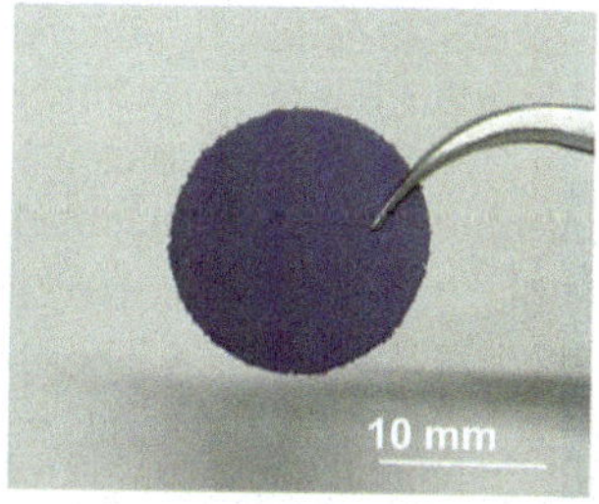

Fig. 3.9 Photograph of the Cu_2(NiTCPP) free-standing membrane

In the in-plane XRD pattern, all the peaks can be assigned to (*hk*0) reflections as a consequence of the highly oriented stacking of nanosheets in the membranes.

In general, crystalline CPs are very fragile due to the three-dimensional network formed by directional coordination and covalent bonds [10d]. However, the present

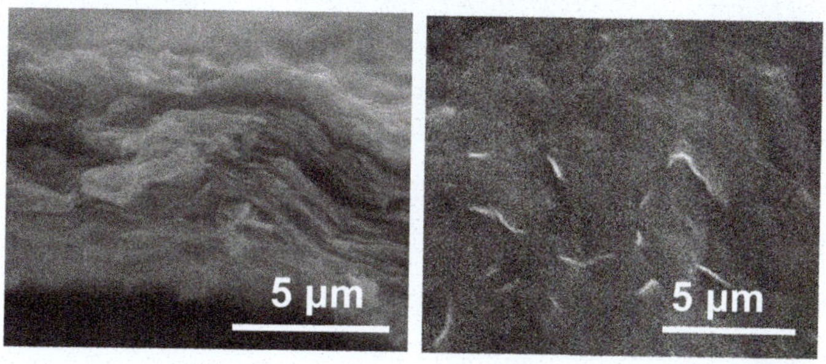

Fig. 3.10 Cross-sectional and surface scanning electron microscope image of the Cu_2(NiTCPP) membrane

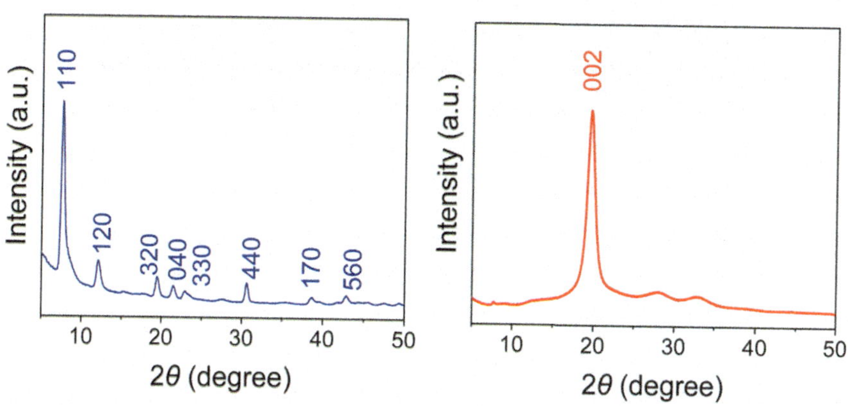

Fig. 3.11 Out-of-plane (red) and in-plane (blue) X-ray diffraction (XRD) patterns of the Cu_2(NiTCPP) membrane

Cu_2(NiTCPP) membrane possesses excellent flexibility, possibly because of the 2D layered structure; the limited bending radius (R) of the 8-μm-thick membrane is less than 1 mm (Figs. 3.12 and 3.14), which is significantly smaller than that of a 500-nm-thick graphene membrane obtained by the chemical vapor deposition technique (R ~ 2.3 mm) [18]. No morphological change was observed on the surface and cross-section of the membrane after the bending and folding operations (Fig. 3.13), demonstrating that the structure of the membrane is intact against the mechanical stress. Furthermore, we confirmed that the membrane returns to the original flat state after repeated bending and relaxing operations of the membrane at least 20 times.

The tensile stress–strain test was performed for the membrane formed in a rectangular shape (Fig. 3.15). As shown in Fig. 3.16, the curves for 13-μm-thick and

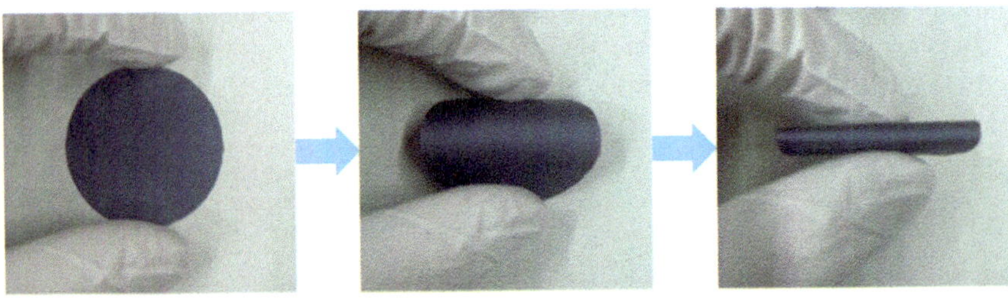

Fig. 3.12 Bending resistance test of the Cu_2(NiTCPP) free-standing membrane with a thickness of 8 μm

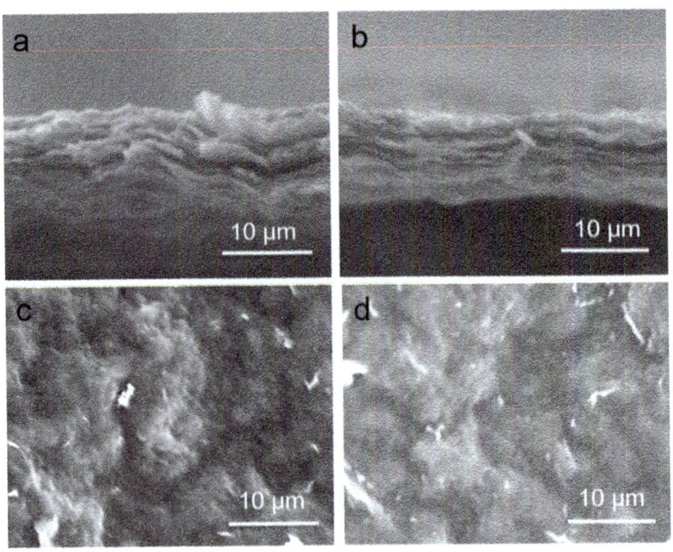

Fig. 3.13 Cross-sectional SEM images of Cu_2(NiTCPP) membrane **a** before and **b** after the bending resistance test, and surface SEM images of Cu_2(NiTCPP) membrane **c** before and **d** after the bending resistance test

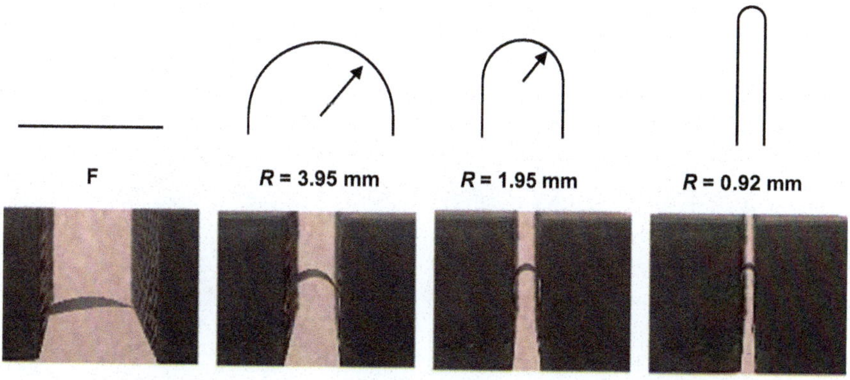

Fig. 3.14 Photographs of the 8-μm-thick membrane during the bending operation

46-μm-thick membranes exhibited a toe region in the initial stage, possibly arising from the stretching of the wrinkled membrane, followed by linear elastic behavior until the abrupt fracture. Young's modulus was calculated to be 1.9 and 0.9 GPa for 13-μm-thick and 46-μm-thick membranes, respectively, from the slope of the curve in the linear region, which is higher than those of commercial Nafion membranes (ca. 0.2 GPa) [19]. It is possible that the higher Young's modulus in the thinner membrane is associated with the less-wrinkle or denser structure in the surface layers of the membranes due to the surface tension during the vacuum filtration. The fracture strength, which is defined as a stress level at the fracture location, is higher in 13-μm-thick membrane (8.6 MPa) than that in 46-μm-thick membrane (6.3 MPa). The reason must be the same as that given for the Young's modulus; namely, the shorter inter-sheet spacing in the thinner membrane [20]. The fracture strain of membranes was estimated to be 0.86% and 0.94% for 13-μm-thick and 46-μm-thick membranes, respectively. It is noticeable that the Cu_2(NiTCPP) membrane exhibited the same level of mechanical properties (namely, Young's modulus, fracture strength, and fracture strain) as membranes of typical 2D nanomaterials such as graphene and WS_2 fabricated by the similar procedure (Table 3.1) [21]. We also performed the three-point bending test of the 46-μm-thick membrane (Fig. 3.17). It should be noted that the flexural modulus (approximately 2.8 GPa) is comparable with typical commercial polymers such as polycarbonate (PC), poly(butylene terephthalate) (PBT), poly(ethylene terephthalate) (PET), and poly(ether ether ketone) (PEEK) [22]. The structural durability of the membranes against the bending and folding operations was verified by measuring in-plane and out-of-plane XRD patterns, which remained unchanged after the bending operation (Fig. 3.18).

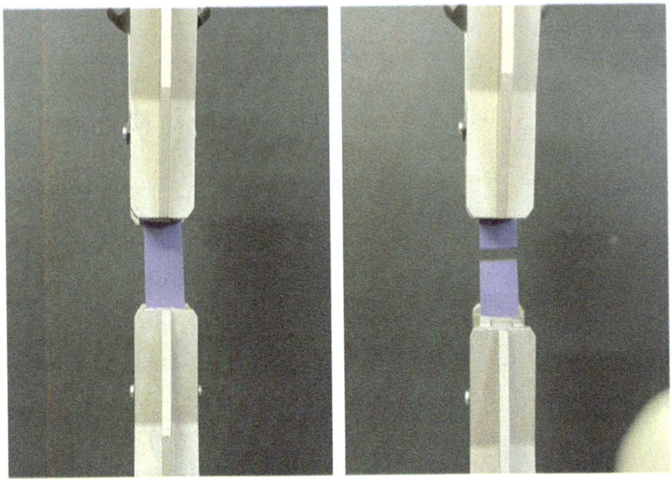

Fig. 3.15 Photographs of the 13-μm-thick membrane before and after the tensile stress–strain test

Fig. 3.16 Photographs of the 13-μm-thick membrane **a** before and **b** after the tensile stress–strain test

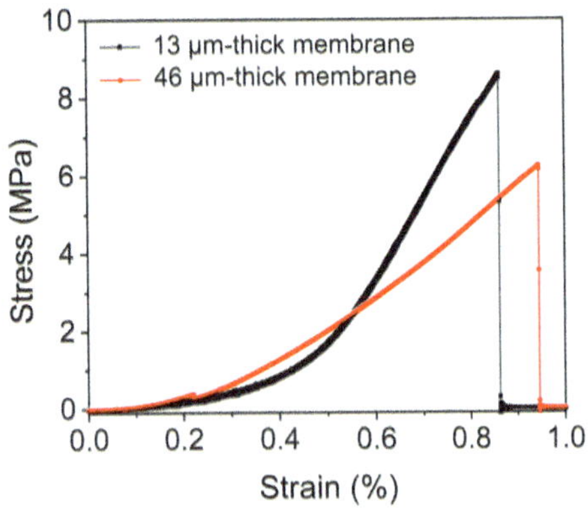

Room-temperature in-plane proton conductivity of the Cu_2(NiTCPP) membrane with a thickness of 60 μm was measured by the a.c. impedance method. As shown in Fig. 3.19, the high-frequency semicircle in the Nyquist plot becomes smaller as the relative humidity (RH) increases from 75% RH to 95% RH. This tendency indicates that the protons originating from the uncoordinated carboxylic groups [23] act as the main charge carrier through the Grotthuss hopping between water molecules in the interlayer spacing. To check the robustness of proton conductivity against the applied bending stress, the impedance spectra of the membrane formed in a rectangular shape (length: 9.40 mm, width: 5.92 mm, thickness: 35 μm) were measured in three kinds of bending states, i.e., flat (F), 1st bending (B1; $R = 5.3$ mm), and 2nd bending (B2; $R = 3.8$ mm). It is obvious that the application of bending stress has little effect on the Nyquist plot (Figs. 3.20 and 3.22), and thereby the proton conductivity (σ)

Table 3.1 Mechanical properties of the Cu_2(NiTCPP) membranes compared with classical material membranes

Samples	Fracture strength (MPa)	Fracture strain (%)	Young's modulus (GPa)	Flexural modulus (GPa)	Ref.
Graphene	6.6	1.2	1.1	/	[21]
BN	< 0.2	0.6	0.2	/	
MoS_2	2.0	0.6	0.4	/	
WS_2	6.8	0.6	2.1	/	
Nafion 117	/	/	~0.2 [5]	0.1–0.4 [6]	[25, 26]
Polycarbonate (PC)	53–72	90–125	2.3–3.0	2.1–2.4	[22]
Poly(butylene terephthalate) (PBT)	52–56	100–300	2.4	1.9–2.6	
Poly(ethylene terephthalate) (PET)	50–70	100	2.7–4.0	2.8	
Poly(ether ether ketone) (PEEK)	70–104	50	3.1–3.8	2.8–3.9	
Cu_2(CuTCPP) (thickness: 50 μm)	4.1	2.5	0.2	/	[13c]
Cu_2(NiTCPP) (thickness: 13 μm)	8.6	0.86	1.9	/	This work
Cu_2(NiTCPP) (thickness: 46 μm)	6.3	0.94	0.9	2.8	

Fig. 3.17 Photographs of the 46-μm-thick membrane during the three-point bending test

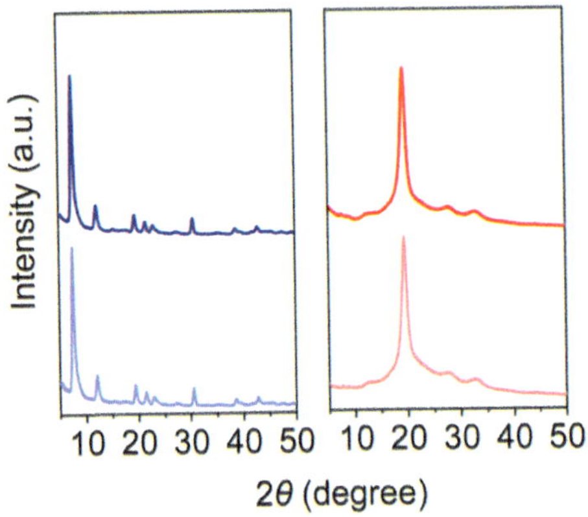

Fig. 3.18 In-plane and out-of-plane XRD patterns of the membrane before (blue/red) and after (pale blue/pale red) bending, as shown in Fig. 3.12

calculated from the semicircle is comparable ranging from 8.0 to 8.5 × 10^{-6} S cm^{-1} regardless of the bending state (Fig. 3.20b). Additional durability test during the repetitive sequences between flat and bending states also affirmed the retention of the proton conduction (Fig. 3.21). Given that the packing structure in the membrane remains intact during the bending operation as shown in Fig. 3.18, the excellent mechanical properties would allow to envisage the present membrane in feasible applications as flexible protonic devices.

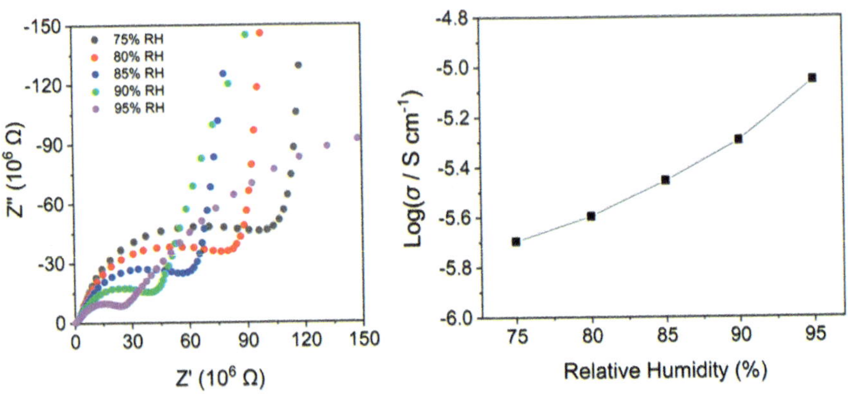

Fig. 3.19 Nyquist plots of the Cu_2(NiTCPP) membrane (length: 8.34 mm, width: 6.13 mm, thickness: 60 μm) at 25 °C under various relative humidity

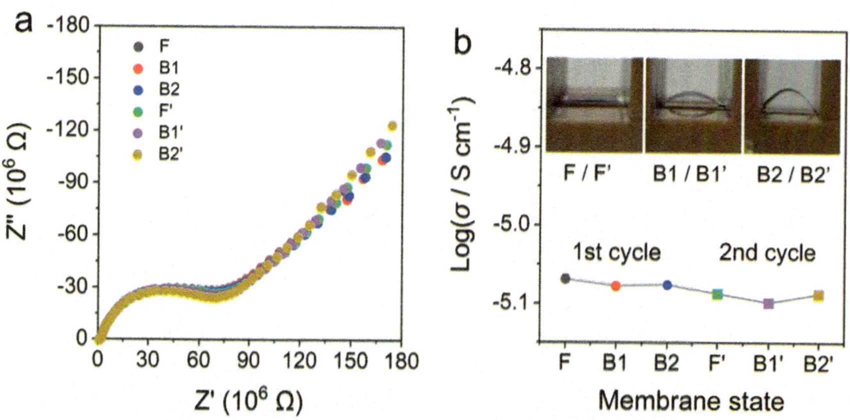

Fig. 3.20 **a** Nyquist plots of the Cu_2(NiTCPP) membrane at 25 °C and 98% RH in various bending states; F: flat, B1: 1st bending state ($R = 5.3$ mm), B2: 2nd bending state ($R = 3.8$ mm); F′, B1′, and B2′ are equivalent to those of F, B1, and B2, respectively, in the second cycle. **b** Proton conductivity (σ) of the Cu_2(NiTCPP) membrane at 25 °C and 98% RH in various bending states (the insets are the photographs of each bending state)

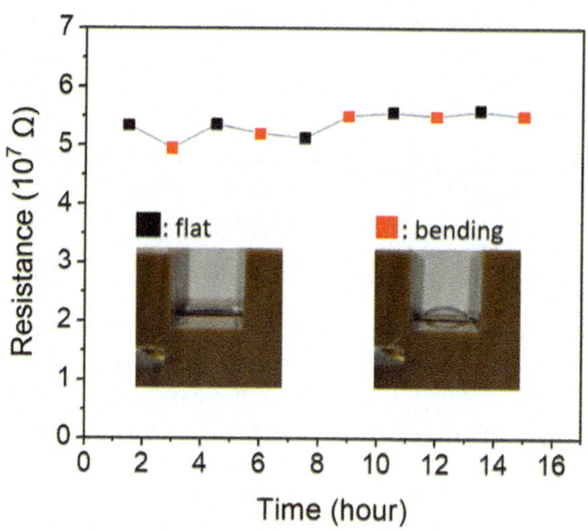

Fig. 3.21 Resistance stability test of the Cu_2(NiTCPP) membrane

Figure 3.23 displays the temperature dependence of Nyquist plots of the membrane in F, B1, and B2 states, respectively. As expected from the above results, the σ value at 25 °C is almost constant with varying bending stress. In all bending states, the σ value steadily increases with increasing temperature and eventually reaches approximately ranging from 1.2 to 1.3 × 10^{-5} S cm^{-1} at 45 °C in different bending states (Fig. 3.24), namely, their proton conductivities are comparable even under heating. Figure 3.25 shows the Arrhenius plots of the product σT of the

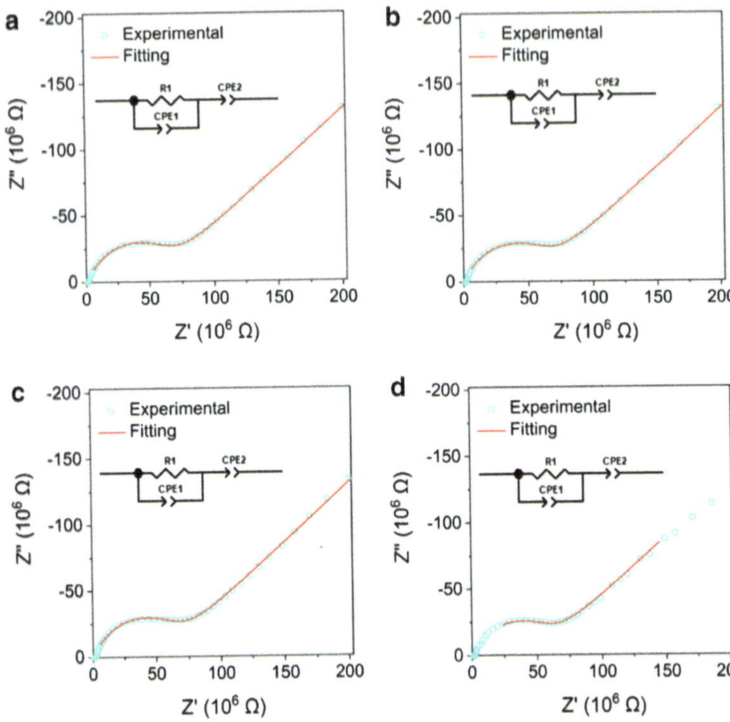

Fig. 3.22 Nyquist plots of the Cu_2(NiTCPP) membrane at 25 °C and 98% RH in various bending states; **a** flat (F), **b** 1st bending (B1), **c** 2nd bending (B2), and **d** flat in the second cycle (F′). Red lines show the fitting curves, where the equivalent circuits used for the simulation are shown in the insets

membrane in the three bending states. Each activation energy (E_a) was estimated to be 0.20, 0.21, and 0.23 eV in F, B1, and B2 states, respectively, using the Arrhenius equation given by $\sigma T = A\ \exp(-E_a/k_BT)$ (A: pre-exponential factor; k_B: Boltzmann constant). The low E_a values suggest that the proton conduction is primarily governed by the Grotthuss mechanism [24], which refers the proton migration by an infinite network of hydrogen bonds in the conducting medium. Given that the E_a value depends largely on the concentration of the proton species or the distance between the proton hopping sites, the comparable values again confirmed the retention of the proton-conducting pathway under applied bending stress.

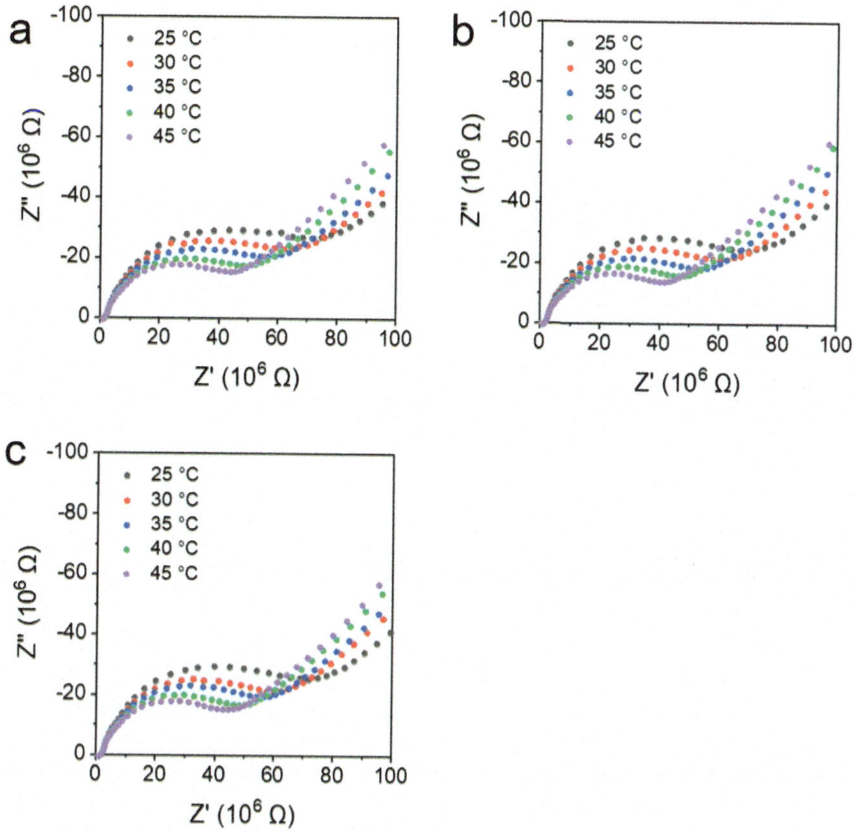

Fig. 3.23 Temperature dependence of the Nyquist plots of the Cu_2(NiTCPP) membrane in the **a** F, **b** B1, and **c** B2 states at 98% RH

Fig. 3.24 Temperature dependence of proton conductivity (σ) of the Cu_2(NiTCPP) membrane in different bending states (F: black, B1: red, B2: blue). Insets are the photographs of each bending state

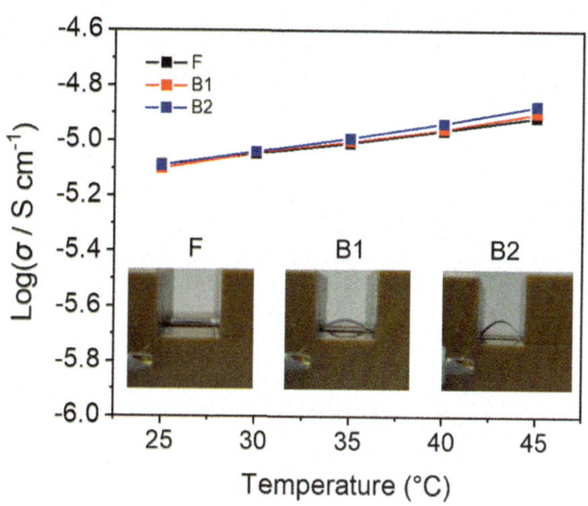

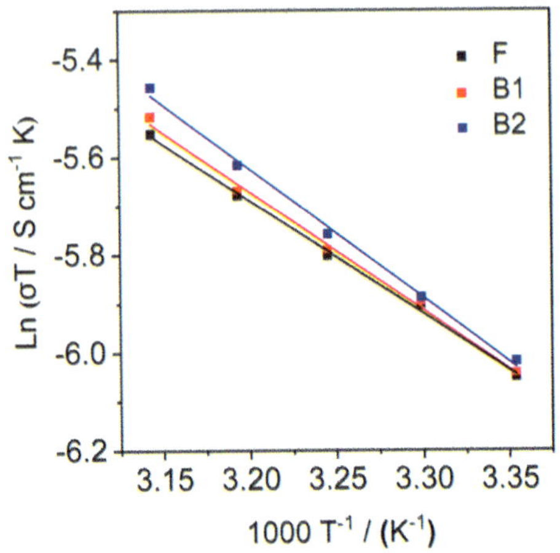

Fig. 3.25 Arrhenius plots of σ of the Cu_2(NiTCPP) membrane in various bending states

3.4 Conclusion

In conclusion, we fabricated highly flexible free-standing membranes composed of 2D CP, Cu_2(NiTCPP). The membrane is highly oriented and excellent mechanical properties by virtue of weak interlayer interactions of van der Waals type. The proton conductivity of the membrane remains almost unchanged even by applying bending stress. The proton-conducting properties, as well as the structural features, in different bending states strongly indicate that proton-conducting pathway through the hydrogen-bonding network is kept intact during the bending operation. This is the first study to demonstrate the robust proton conduction of 2D CP against the applied bending stress. The prominent advantage of the present membrane for future flexible protonic devices is that it can be fabricated free from any substrate and shows excellent bending stability even without blending with polymer matrix (Fig. 3.26). Our present study will offer an opportunity to extend the range of feasible applications of CPs in protonic devices.

Fig. 3.26 Nyquist plots of the flat Cu_2(NiTCPP) membrane at 25 °C and 98% RH before (black) and after (red and blue) the bending operation

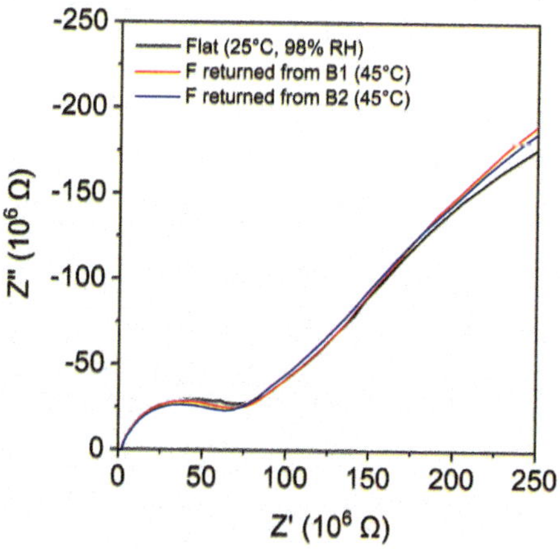

References

1. (a) O.M. Yaghi, M. O'Keeffe, N.W. Ockwig, H.K. Chae, M. Eddaoudi, J. Kim, Nature **423,** 705–714 (2003); (b) S. Kitagawa, R. Kitaura, S.I. Noro, Angew. Chem. Int. Ed. **43,** 2334–2375 (2004), Angew. Chem. **43,** 2334–2375 (2004); (c) H.C. Zhou, J.R. Long, O.M. Yaghi, Chem. Rev. **112,** 673–674 (2012)
2. J. Canivet, A. Fateeva, Y. Guo, B. Coasne, D. Farrusseng, Chem. Soc. Rev. **43**, 5594–5617 (2014)
3. (a) Y. Ye, L. Gong, S. Xiang, Z. Zhang, B. Chen, Adv. Mater. **32,** 1907090 (2020); (b) S. Paul, S.J. Choi, H.J. Kim, Energy Fuels **34,** 10067–10077 (2020)
4. G.D. Wu, H.L. Zhou, Z.H. Fu, W.H. Li, J.W. Xiu, M.S. Yao, Q.H. Li, G. Xu, Angew. Chem. Int. Ed. **60,** 9931–9935 (2021), Angew. Chem. **133,** 10019–10023 (2021)
5. (a) S.Z. Huang, S.S. Liu, H.J. Zhang, Z. Han, G. Zhao, X.Y. Dong, S.Q. Zang, ACS Appl. Mater. Interfaces **12,** 28720–28726 (2020); (b) L.D. Costanzo, B. Panunzi, Molecules **26,** 2952 (2021); (c) S.L. Yang, G. Li, M.Y. Guo, W.S. Liu, R. Bu, E.Q. Gao, J. Am. Chem. Soc. **143,** 8838–8848 (2021)
6. (a) X. Li, H. Zhang, J. Hou, R. Ou, Y. Zhu, C. Zhao, T.Y. Qian, C.D. Easton, C. Selomulya, M.R. Hill, H. Wang, J. Am. Chem. Soc. **142,** 9827–9833 (2020); (b) J. Lu, H. Xu, H. Yu, X. Hu, J. Xia, Y. Zhu, F.C. Wang, H.A. Wu, L. Jiang, H. Wang, Sci. Adv. **8,** eabl5070 (2022)
7. (a) P. Ramaswamy, N.E. Wong, G.K. Shimizu, Chem. Soc. Rev. **43,** 5913–5932 (2014); (b) Y. Ye, L. Gong, S. Xiang, Z. Zhang, B. Chen, Adv. Mater. **32,** 1907090 (2020); (c) D.W. Lim, H. Kitagawa, Chem. Soc. Rev. **50,** 6349–6368 (2021)
8. H.S. Sasmal, H.B. Aiyappa, S.N. Bhange, S. Karak, A. Halder, S. Kurungot, R. Banerjee, Angew. Chem. Int. Ed. **57,** 10894–10898 (2018), Angew. Chem. **130,** 11060–11064 (2018)
9. (a) S. Kusumoto, Y. Kim, S. Hayami, Coord. Chem. Rev. **475**, 214890; (b) X. Zhou, H. Jin, B.Y. Xia, K. Davey, Y. Zheng, S.Z. Qiao, Adv. Mater. **33,** 2104341 (2021)
10. (a) M. Đaković, M. Borovina, M. Pisačić, C.B. Aakeröy, Ž. Soldin, B.M. Kukovec, I. Kodrin, Angew. Chem. Int. Ed. **57,** 14801–14805 (2018), Angew. Chem. **130,** 15017–15021 (2018); (b) M. Pisacic, I. Biljan, I. Kodrin, N. Popov, Z. Soldin, M. Đaković, Chem. Mater. **33,** 3660–3668 (2021); (c) B. Bhattacharya, A.A. Michalchuk, D. Silbernagl, M. Rautenberg, T. Schmid, T. Feiler, K. Reimann, A. Ghalgaoui, H. Sturm, B. Paulus, F. Emmerling, Angew. Chem. Int. Ed. **59,** 5557–5561 (2020), Angew. Chem. **59,** 5557–5561 (2020); (d) L.C. An, X. Li, Z.G. Li, Q. Li, P.J. Beldon, F.F. Gao, Z.Y. Li, S. Zhu, L. Di, S. Zhao, J. Zhu, D. Comboni, I. Kupenko, W. Li, U. Ramamurty, X.H. Bu, Nat. Commun. **13,** 6645 (2022)

11. (a) J. Yan, T. Liu, X. Liu, Y. Yan, Y. Huang, Coord. Chem. Rev. **452,** 214300 (2022); (b) Y. Shi, B. Liang, R.B. Lin, C. Zhang, B. Chen, Trends Chem. **2,** 254–269 (2020); (c) F. Jin, F. J. Liu, Y. Chen, Z. Zhang, Angew. Chem. Int. Ed. **60,** 14222–14235 (2021), Angew. Chem. **133,** 14342–14355 (2021)
12. (a) O. Shekhah, J. Liu, R.A. Fischer, C. Wöll, Chem. Soc. Rev. **40,** 1081–1106 (2011); (b) J. Dechnik, J. Gascon, C.J. Doonan, C. Janiak, C.J. Sumby, Angew. Chem. Int. Ed. **56,** 9292–9310 (2017), Angew. Chem. **129,** 9420–9439 (2017)
13. (a) G. Xu, T. Yamada, K. Otsubo, S. Sakaida, H. Kitagawa, J. Am. Chem. Soc. **134,** 16524–16527 (2012); (b) G. Xu, K. Otsubo, T. Yamada, S. Sakaida, H. Kitagawa, J. Am. Chem. Soc. **135,** 7438–7441 (2013); (c) M. Tian, F. Pei, M. Yao, Z. Fu, L. Lin, G. Wu, G. Xu, H. Kitagawa, X. Fang, Energy Storage Mater. **21,** 14–21 (2019); (d) J. Lu, Y. Yoshida, M. Maesato, H. Kitagawa, Angew. Chem. Int. Ed. **61,** e202213077 (2022), Angew. Chem. **134,** e202213077 (2022)
14. (a) G. Granados-Oliveros, E.A. Páez-Mozo, F.M. Ortega, C. Ferronato, J.M. Chovelon, Appl. Catal. B **89,** 448–454 (2009); (b) P. K. Sonkar, K. Prakash, M. Yadav, V. Ganesan, M. Sankar, R. Gupta, D.K. Yadav, J. Mater. Chem. **5,** 6263–6276 (2017); (c) K. Jeong, J. M. Kim, S.H. Kim, G.Y. Jung, J. Yoo, S.H. Kim, S.K. Kwak, S.Y. Lee, Adv. Funct. Mater. **29,** 1806937 (2019)
15. R. Makiura, S. Motoyama, Y. Umemura, H. Yamanaka, O. Sakata, H. Kitagawa, Nat. Mater. **9,** 565–571 (2010)
16. L. Wang, P. Jin, J. Huang, H. She, Q. Wang, ACS Sustain. Chem. Eng. **7,** 15660–15670 (2019)
17. C. Berrios, G.I. Cardenas-Jiron, J.F. Marco, C. Gutierrez, M.S. Ureta-Zanartu, J. Phys. Chem. A **111,** 2706–2714 (2007)
18. J. Yun, D. Kim, G. Lee, J.S. Ha, Carbon **79,** 156–164 (2014)
19. M.B. Satterfield, J.B. Benziger, J. Polym. Sci. B Polym. Phys. **47,** 11–24 (2009)
20. (a) A.A. Griffith, Philos. Trans. R. Soc. London A **221,** 163 (1921); (b) T. Gong, D.V. Lam, R. Liu, S. Won, Y. Hwangbo, S. Kwon, J. Kim, K. Sun, J.H. Kim, S.M. Lee, C. Lee, Adv. Funct. Mater. **25,** 3756–3763 (2015)
21. J.N. Coleman, M. Lotya, A. O'Neill, S.D. Bergin, P.J. King, U. Khan, K. Young, A. Gaucher, S. De, R.J. Smith, I.V. Shvets, S.K. Arora, G. Stanton, H.Y. Kim, K. Lee, G.T. Kim, G.S. Duesberg, T. Hallam, J.J. Boland, J.J. Wang, J.F. Donegan, J.C. Grunlan, G. Moriarty, A. Shmeliov, R.J. Nicholls, J.M. Perkins, E.M. Grieveson, K. Theuwissen, D.W. McComb, P.D. Nellist, V. Nicolosi, Science **331,** 568–571 (2011)
22. K. Van de Velde, P. Kiekens, Polym. Test. **20,** 885–893 (2001)
23. Y.N. Zhou, L.L. Liu, Q.W. Liu, X.X. Liu, M.Z. Feng, L. Wang, Z.G. Sun, Y.Y. Zhu, X. Zhang, C.Q. Jiao, Inorg. Chem. **60,** 17303–17314 (2021)
24. D. Marx, ChemPhysChem **7,** 1848–1870 (2006)
25. M.B. Satterfield, J.B. Benziger, J. Polym. Sci. B: Polym. Phys. **47,** 11–24 (2009)
26. Y. Wang, H. Chen, Y. Wang, B. Luo, L. Chang, Z. Zhu, B. Li, Polym. Eng. Sci. **54,** 818–830 (2014)

Chapter 4
Proton-Conducting Behavior in Flexible Coordination Polymer Free-Standing Membranes

Abstract Applying CP to flexible devices remains a considerable challenge because of its brittleness. CP structure of low dimension usually plays a crucial role in deciding on the mechanical flexibility, typical dimensional coordination chain that acts as a rigid frame and the other two directions through many weak molecular interactions. We employed a series layered CP, Cu_2(MTCPP) (H_4(H_2TCPP); 5,10,15,20-tetrakis(4-carboxyphenyl)porphyrin) (M = Ni(II), Cu(II), Zn(II)), in which a two-dimensional (2D) square grid sheet composed of tetradentate metal porphyrins and paddlewheel-type copper dimers is connected by weak van der Waals forces. These membranes composed with two-dimensional (2D) nanosheets show good orientation and bending tests evaluated their mechanical flexibility. Electrochemical impedance spectroscopy revealed that the in-plane proton conductivities of the membranes show different conductivities due to the difference in the hydration coordination ability of central porphyrin metals. Our present study provides a promising strategy for fabricating new and more advanced 2D CPs without using substrates and additional polymers for photonic devices.

Keywords Two-dimensional (2D) · Nanosheets · Orientation · Flexible membrane · Proton conducting

4.1 Introduction

As an emerging porous material, porous coordination polymer (PCP) nanosheets have attracted more and more attention due to their advantages, such as sizeable transverse size, ultra-thin thickness, customizable physical and chemical environment, flexibility, and easy access to active surface sites [1]. Due to the ultra-high designability of PCP, its pore size can be adjusted [2] and its structure can store more water molecules [3]. Water molecules form hydrogen bond networks inside or on the surface of PCP, which can effectively provide proton conduction pathways [4]. Therefore, PCP has become a candidate for proton conductor, which has aroused

J. Lu, *Novel Proton Conducting in Free-Standing Coordination Polymer Membranes*, Springer Theses, https://doi.org/10.1007/978-981-97-7913-0_4

the interest of many researchers. However, their single crystals composed of directional coordination bonds are inherently brittle, severely limiting their application in flexible electronics and intelligent wearable devices [5].

How to apply the CPs to wearable devices and flexible electronics field, the CPs preparation membrane will maximize its role [6]. Currently, there are two main ways to prepare CPs membranes. The first is to grow CPs on the substrate [7]. CPs are composed of directional coordination bonds and are brittle. The second is to embed CPs into the polymer matrix [8], and the thickness of the polymer matrix is much larger than the diameter of the porous CPs. Thus, the doped CPs are surrounded by a polymer matrix. In this case, the properties of CPs/polymer hybrid membranes significantly depend on the polymer properties. So far, no flexible PCP free-standing membrane of the system has been reported. In theory, pure CPs membranes provide the best performance [9]. However, their structural fragility limits their application in flexible electronics and intelligent wearable devices. How to obtain pure CPs films with flexibility is a significant obstacle to the practical application of CPs. Therefore, the immediate challenge is to design an independent, flexible membrane composed of CPs nanosheets without losing its structural characteristics such as crystallinity and porosity [10]. The low-dimensional CP structure often plays a crucial role in determining flexibility. Typically, a common-dimensional coordination chain serves as a rigid frame, while the other two directions are stabilized through numerous weak molecular interactions [11].

$Cu_2(CuTCPP)$ ($H_4(H_2TCPP)$; 5,10,15,20-tetrakis(4-carboxyphenyl)porphyrin) membrane was first reported by Xu et al. [12], whereas our group recently realized an all-solid proton-rectifying device with exceptionally high performance (rectification ratio > 200) using $Cu_2(CuTCPP)$ membrane [13]. In this study, we prepared a series of Cu(II) the paddle wheel unit and four teeth H_4MTCPP (($H_4(H_2TCPP)$; 5,10,15,20-tetrakis(4-carboxyphenyl)porphyrin); M = Ni(II), Cu(II), Zn(II)) ligand. These membranes show good orientation, and its mechanical flexibility was evaluated by bending tests; electrochemical impedance spectroscopy revealed that the in-plane proton conductivities of the membranes show different conductivities due to the difference in the hydration coordination ability of porphyrin central metals. Our present study provides a promising strategy for fabricating new and more advanced 2D CPs without using any substrates and addition polymers for protonic devices.

4.2 Experimental

4.2.1 *Materials*

Copper(II) nitrate trihydrate (99%), copper(II) acetate monohydrate (99%), nickel(II) acetate tetrahydrate (99%), zinc(II) acetate dehydrate (99%), and *N*,*N*-dimethylformamide (DMF, 99.5%) were used as received from Fujifilm Wako Pure

Chemical Industries, Ltd. Tetrakis(4-carboxyphenyl)porphyrin ($H_4(H_2TCPP)$, 97%) was purchased from Tokyo Chemical Industry Co., Ltd. and used without further purification. Water was deionized before use.

4.2.2 Characterizations

Powder X-ray diffraction (PXRD) measurements were performed using a Bruker D8 ADVANCE with Cu Kα radiation ($\lambda = 0.15418$ nm) in the 2θ range from 5° to 60° with a 0.02° step and a step time of 1 s. In-plane and out-of-plane grazing-incidence XRD patterns were collected using a Rigaku Smart-Lab X-ray diffractometer with Cu Kα radiation ($\lambda = 0.15418$ nm) in the 2θ range from 5 to 50° at a scanning rate of 0.5° min^{-1}. Fourier-transform infrared (FTIR) spectra were obtained using a PerkinElmer Spectrum 100 spectrometer in attenuated total reflection (ATR) mode. Thermal stability was characterized by thermogravimetric (TG) measurements using a Bruker TG-DTA2000SA instrument at a heating rate of 5 °C min^{-1} under flowing N_2. The sorption experiments for N_2 (77 K) was performed using a BELSORP-max apparatus (MicrotracBEL). Ultraviolet–visible (UV–Vis) absorption spectra were measured using a JASCO V-570 UV–Vis-NIR spectrometer in a NaOH solution. Scanning electron microscope (SEM) images of the nanosheets and membranes were obtained using a Hitachi SU-1510 instrument. Transmission electron microscope (TEM) images of the nanosheets were obtained using a Hitachi HT7700 instrument. The strain–stress and flexural modulus experiments were performed by a Autograph AG-Xplus instrument. The impedance spectra of the membranes were measured using an Impedance/Gain-Phase Analyzer SI 1260 at an applied a.c. voltage of 100 mV. Gold paste (Tokuriki, 8560-1A) was painted on opposite sides of the membrane.

4.2.3 Preparation of H_4MTCPP Ligands

4.2.3.1 Preparation of H_4NiTCPP Ligands

H_4NiTCPP ligand was prepared by refluxing $H_2(H_4TCPP)$ (0.33 mmol) with nickel(II) acetate tetrahydrate (1.82 mmol) in N,N′-dimethylformamide (70 mL) for 2 h. DMF was removed by distillation, and the crude H_4(NiTCPP) was precipitated by adding water, washed with water, and then collected by centrifugation. The precipitate was dissolved in 100 mL 0.1 M NaOH solution and recrystallized by adding 20 mL 1 M HCl solution. Recrystallized sample was washed by water and collected by vacuum filtration. The precipitate was dissolved in ethanol and recrystallized by solvent evaporation.

4.2.3.2 Preparation of $H_4CuTCPP$ Ligands

H_4MTCPP ligand was prepared by refluxing $H_2(H_4TCPP)$ (0.33 mmol) with copper(II) acetate monohydrate (1.82 mmol) in N,N′-dimethylformamide (70 mL) for 6 h. DMF was removed by distillation, and the crude $H_4(NiTCPP)$ was precipitated by adding water, washed with water, and then collected by centrifugation. The precipitate was dissolved in 100 mL 0.1 M NaOH solution and recrystallized by adding 20 mL 1 M HCl solution. Recrystallized sample was washed by water and collected by vacuum filtration. The precipitate was dissolved in ethanol and recrystallized by solvent evaporation.

4.2.3.3 Preparation of $H_4ZnTCPP$ Ligands

$H_4ZnTCPP$ ligand was prepared by refluxing $H_2(H_4TCPP)$ (0.33 mmol) with zinc(II) acetate dehydrate (1.82 mmol) in N,N′-dimethylformamide (70 mL) for 6 h. DMF was removed by distillation, the crude $H_4(NiTCPP)$ was precipitated by adding water, washed with water, and then collected by centrifugation, and the H4MTCPPs were precipitated by adding water and washing with water and then collected by centrifuge. The precipitate was dissolved in 100 mL 0.1 M NaOH solution and recrystallized by adding 20 mL 1 M HCl solution. Recrystallized sample was washed by water and collected by vacuum filtration. The precipitate was dissolved in ethanol and recrystallized by solvent evaporation.

4.2.4 *Preparation of Cu_2MTCPP Nanosheets*

4.2.4.1 Preparation of $Cu_2NiTCPP$ Nanosheets

About 8.5 mg (0.01 mmol) of $H_4NiTCPP$ was dissolved into 13 mL of DMF, then adding 5 mL 1 mg/L $Cu(NO_3)_2·3H_2O$ DMF (0.02 mmol) solution and ultrasonicated for 10 min. The mixture was aged at 35 °C for 10 h and then heated at 85 °C for 24 h. The mixture was cooled to room temperature within 12 h. Finally, the gel product was collected by centrifugation (7830 rpm for 10 min) and washed with ethanol more than five times until the supernatant became colorless. The purple polycrystalline product was then dried at 50 °C for 12 h (8.0 mg, yield (ca. 81.8%)).

4.2.4.2 Preparation of $Cu_2CuTCPP$ Nanosheets

About 8.5 mg (0.01 mmol) of $H_4CuTCPP$ was dissolved into 13 mL of DMF, then adding 5 mL 1 mg/L $Cu(NO_3)_2·3H_2O$ (0.02 mmol) solution and ultrasonicated for 10 min. The mixture was aged at 35 °C for 10 h and then heated at 85 °C for 24 h. The mixture was cooled to room temperature within 12 h. Finally, the gel product was collected by centrifugation (7830 rpm for 10 min) and washed with ethanol more than five times until the supernatant became colorless. The purple polycrystalline product was then dried at 50 °C for 12 h (6.8 mg, yield (ca. 69.5%)).

4.2.4.3 Preparation of $Cu_2ZnTCPP$ Nanosheets

About 8.5 mg (0.01 mmol) of $H_4ZnTCPP$ was dissolved into 13 mL of DMF, then adding 5 mL 1 mg/L $Cu(NO_3)_2 \cdot 3H_2O$ (0.02 mmol) solution and ultrasonicated for 10 min. The mixture was aged at 35 °C for 10 h and then heated at 85 °C for 24 h. The mixture was cooled to room temperature within 12 h. Finally, the gel product was collected by centrifugation (7830 rpm for 10 min) and washed with ethanol more than five times until the supernatant became colorless. The purple polycrystalline product was then dried at 50 °C for 12 h (7.4 mg, yield (ca. 75.7%)).

4.2.5 Preparation of Cu_2MTCPP Membranes

4.2.5.1 Preparation of $Cu_2NiTCPP$ Membranes

The $Cu_2NiTCPP$ gel product was added to 200 mL of ethanol in a 300 mL flask, which was tightly sealed after purging with nitrogen gas. Ultrasonic treatment for 90 min allowed the exfoliated nanosheets to disperse well in ethanol. A $Cu_2NiTCPP$ free-standing membrane was prepared by filtering the ethanol solution in which the $Cu_2NiTCPP$ nanosheets were dispersed. Briefly, 50 mL of the ethanolic $Cu_2NiTCPP$ suspension was vacuum-filtered through a polyvinylidene difluoride (PVDF) membrane filter (pore diameter: ~0.22 μm; porosity: ~80%; diameter: 25 mm) and dried at 50 °C for 4 h. The $Cu_2NiTCPP$ membrane, with a thickness of approximately 8 μm, could be easily detached from the filter to allow free-standing operation. The membrane with 34 mm diameter was fabricated by the same method using a setup of diameter of 45 mm.

4.2.5.2 Preparation of $Cu_2CuTCPP$ Membranes

The $Cu_2CuTCPP$ gel product was added to 200 mL of ethanol in a 300 mL flask, which was tightly sealed after purging with nitrogen gas. Ultrasonic treatment for 90 min allowed the exfoliated nanosheets to disperse well in ethanol. A $Cu_2CuTCPP$ free-standing membrane was prepared by filtering the ethanol solution in which the $Cu_2CuTCPP$ nanosheets were dispersed. Briefly, 50 mL of the ethanolic $Cu_2CuTCPP$ suspension was vacuum-filtered through a polyvinylidene difluoride (PVDF) membrane filter (pore diameter: ~0.22 μm; porosity: ~80%; diameter: 25 mm) and dried at 50 °C for 4 h. The $Cu_2CuTCPP$ membrane, with a thickness of approximately 8 μm, could be easily detached from the filter to allow free-standing operation. The membrane with 34 mm diameter was fabricated by the same method using a setup of diameter of 45 mm.

4.2.5.3 Preparation of $Cu_2ZnTCPP$ Membranes

The Cu_2(ZnTCPP) gel product was added to 200 mL of ethanol in a 300 mL flask, which was tightly sealed after purging with nitrogen gas. Ultrasonic treatment for 90 min allowed the exfoliated nanosheets to disperse well in ethanol. A Cu_2(ZnTCPP) free-standing membrane was prepared by filtering the ethanol solution in which the Cu_2(ZnTCPP) nanosheets were dispersed. Briefly, 50 mL of the ethanolic Cu_2(ZnTCPP) suspension was vacuum-filtered through a polyvinylidene difluoride (PVDF) membrane filter (pore diameter: ~0.22 μm; porosity: ~80%; diameter: 25 mm) and dried at 50 °C for 4 h. The Cu_2(ZnTCPP) membrane, with a thickness of approximately 8 μm, could be easily detached from the filter to allow free-standing operation. The membrane with 34 mm diameter was fabricated by the same method using a setup of diameter of 45 mm.

4.3 Results and Discussion

Cu_2(MTCPP) (M: Ni/Cu/Zn) nanosheets were prepared by a stepwise synthesis procedure [14]. H_4(MTCPP) was prepared by reacting *meso*-tetrakis(4-carboxyphenyl)porphyrin ($H_4(H_2TCPP)$) with $Ni(OAc)_2{\cdot}4H_2O/Cu(OAc)_2{\cdot}H_2O/Zn(OAc)_2{\cdot}2H_2O$, respectively, according to literature (Detailed see Experiment Part) [14]. The successful synthesis of H_4(MTCPP) was verified by ultraviolet–visible (UV–vis) absorption, as shown in Fig. 4.1, The number of Q band peaks decreases from three in $H_4(H_2TCPP)$ to one or two in H_4(MTCPP), which demonstrates that the porphyrin center in H_4(NiTCPP) was filled with a nickel(II)/copper(II)/zinc(II) ions. Fourier-transform infrared (FTIR) measurement was also used to prove the coordination between $H_4(H_2TCPP)$ unit and metal ions. In Fig. 4.2, a N–H stretching band (3319 cm^{-1}), which is found in $H_4(H_2TCPP)$, disappears in Cu_2(MTCPP). The number of peaks at around 1000 cm^{-1} decreases from four in $H_4(H_2TCPP)$ (N–H bending vibration) to one in H_4(MTCPP) (N–M stretching vibration), which demonstrates that the porphyrin center in H_4(NiTCPP) has been filled with nickel(II)/copper(II)/zinc(II) ions, respectively. In addition, ^1H nuclear magnetic resonance (NMR) spectra (Fig. 4.3) were used to further prove the successful synthesis of H_4(MTCPP). ^1H NMR (600 MHz, DMSO-d_6) H_4(NiTCPP): δ 13.28 (s, 4H), 8.76 (s, 8H), 8.30 (d, 8H), 8.14 (d, 8H) ppm; H_4(ZnTCPP): δ 13.26 (s, 4H), 8.81 (s, 8H), 8.37 (d, 8H), 8.31 (d, 8H) ppm.

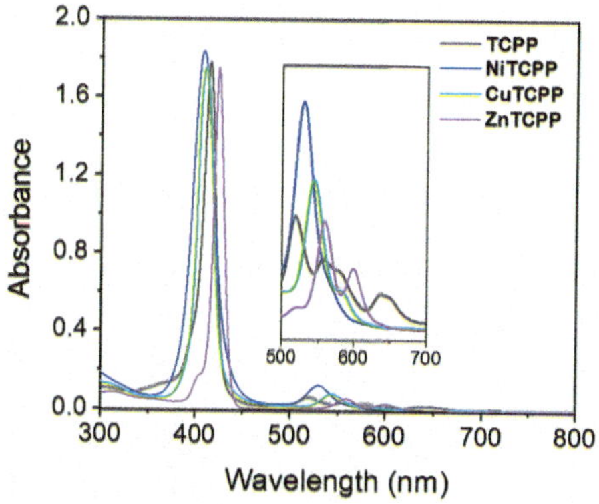

Fig. 4.1 UV–vis spectra of $H_4(H_2TCPP)$ (black line) and $H_4(NiTCPP)$ (blue line)/ $H_4(CuTCPP)$ (green line)/ $H_4(ZnTCPP)$ (purple line) in a NaOH solution

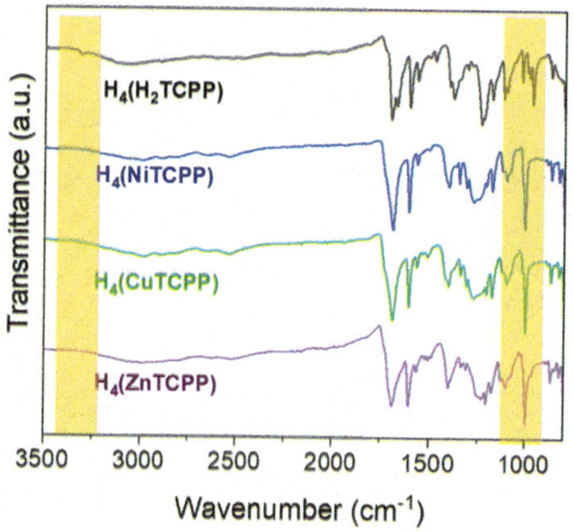

Fig. 4.2 FTIR spectra of $H_4(H_2TCPP)$ (black line) and $H_4(NiTCPP)$ (blue line)/ $H_4(CuTCPP)$ (green line)/ $H_4(ZnTCPP)$ (purple line)

Purple crystalline $Cu_2(MTCPP)$ nanosheets were obtained by a solvothermal reaction in *N*,*N*-dimethylformamide, used as a solvent, as previously reported (Detailed see Experiment part) [12]. The powder X-ray diffraction (XRD) pattern of the pulverized $Cu_2(MTCPP)$ nanosheets is consistent with the simulated pattern of homobimetallic 2D CP, Cu $_2$(CuTCPP) reported previously (Fig. 4.4), with a layered structure [15]. As shown in Fig. 4.5, each $Cu_2(MTCPP)$ sheet has a square grid structure in the *ab* plane, which is composed of tetradentate metal(II) porphyrin ($MTCPP^{4-}$) and paddlewheel-type copper(II) dimers. Along the *c*-axis, the reticulate layers were assembled mainly through the van der Waals interactions in the AB stacking mode, where the B layer shifted by approximately *a*/4 relative to the A layer. The FTIR spectra show the disappearance of the peak corresponding to the N–H bending mode (966 cm^{-1} in $H_4(H_2TCPP)$) in $H_4(MTCPP)$, indicating the metallization of the porphyrin units (Fig. 4.6). The disappearance of the C=O stretching mode in COOH groups (1676 cm^{-1} in $H_4(NiTCPP)$) in $Cu_2(NiTCPP)$ confirms that four carboxyl groups of the MTCPP units successfully coordinated to the adjacent copper ions to form copper paddlewheel units [15]. In the XPS survey spectra of the

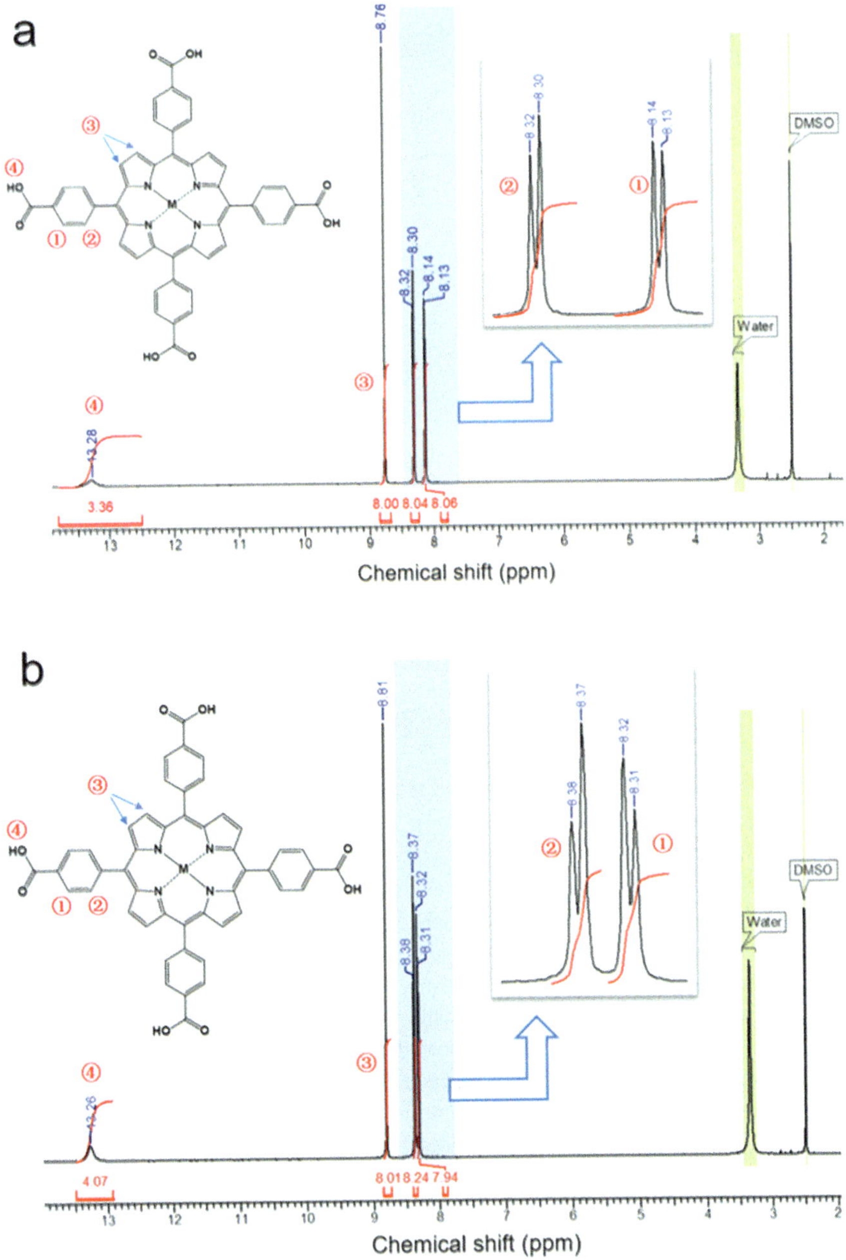

Fig. 4.3 ^{1}H NMR spectra of **a** H_4(NiTCPP) and **b** H_4(ZnTCPP) in DMSO-d_6

obtained 2D CPs, in the CPs possessing bimetallic active centers, both of the two metal elements were found. In the high-resolution XPS spectra of Ni 2p, Cu 2p, Zn 2p (Fig. 4.7), the peaks located at about 872 and 855 eV belong to Ni $2p_{1/2}$ and Ni $2p_{3/2}$ [16], 955 and 935 eV belong to Cu $2p_{1/2}$ and Cu $2p_{3/2}$ [17], and 1045 and 1022 eV belong to Zn $2p_{1/2}$ and Zn $2p_{3/2}$ [18], respectively. In Cu_2(CuTCPP), both Cu 2p 3/2 and 2p 1/2 spin–orbit peaks involve two distinct components as a reflection of two kinds of coordination environments [19]; one being in a paddlewheel-type dimer coordinated by four oxygens and another being in a porphyrin macrocycle coordinated by four nitrogens. The transmission electron microscope (TEM) image of Cu_2(MTCPP) deposited onto a carbon-coated copper substrate by dropping ethanol suspension reveals a nanosheet-like morphology (Fig. 4.8).

Owing to the good dispersion of Cu_2(MTCPP) nanosheets in ethanol, the ethanol suspension was filtered through a polyvinylidene difluoride (PVDF) membrane filter to yield a highly oriented membrane. The membrane with high flexibility could be readily peeled from the filter after drying at 50 °C. Figure 4.9 shows a photograph of the free-standing membrane with a diameter of 17 mm and a thickness of approximately 8 μm. The thickness can be controlled in the range of less than ten to hundreds of micrometers by adjusting the loading amounts of the suspension. The cross-sectional scanning electron microscope (SEM) image (Fig. 4.10) shows the membrane formation by stacking nanosheets, which is consistent with the SEM image of the membrane surface. Figure 4.11 shows the out-of-plane and in-plane grazing-incidence XRD patterns of the membranes. The out-of-plane XRD pattern was dominated by the interlayer (002) reflection, indicating the preferred orientation of the membrane along the crystallographic *c*-axis. In the in-plane XRD pattern, all the peaks can be assigned to (*hk*0) reflections because of the highly oriented stacking of nanosheets in the membranes. Crystalline CPs are highly fragile because of their three-dimensional network formed by directional coordination and covalent bonds [5]. However, the present Cu_2(NiTCPP) membrane possessed excellent flexibility,

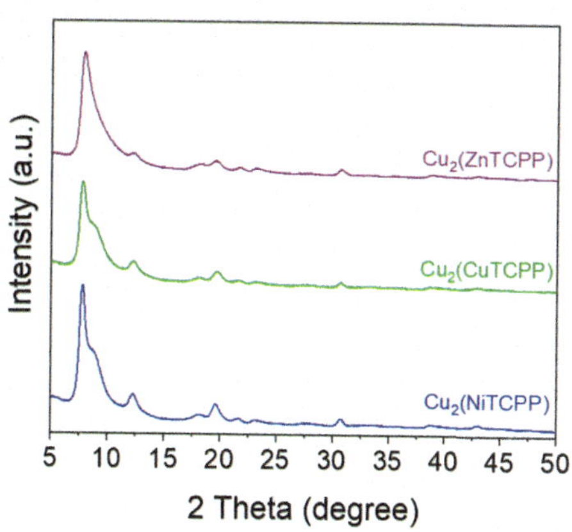

Fig. 4.4 PXRD pattern of pulverized Cu_2(NiTCPP) nanosheets (blue), Cu_2(NiTCPP) nanosheets (green) and Cu_2(NiTCPP) nanosheets (purple)

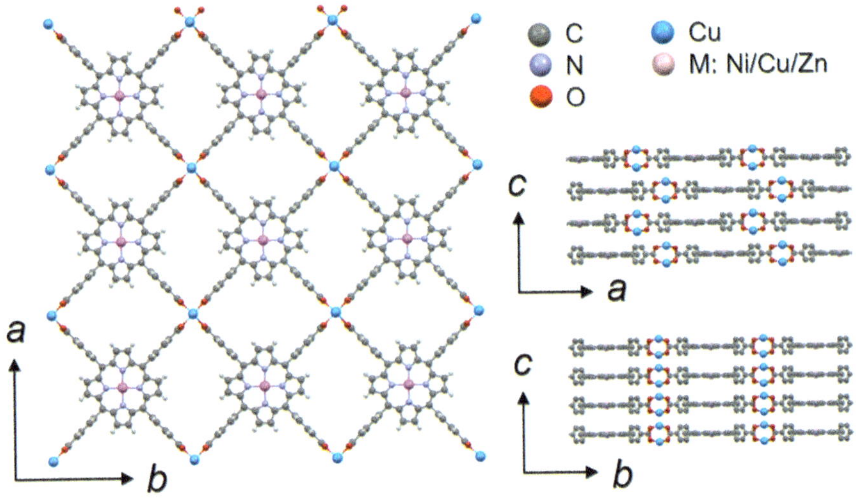

Fig. 4.5 **a** Simulated 2D sheet structure of Cu_2(MTCPP) (M: Ni/Cu/Zn)

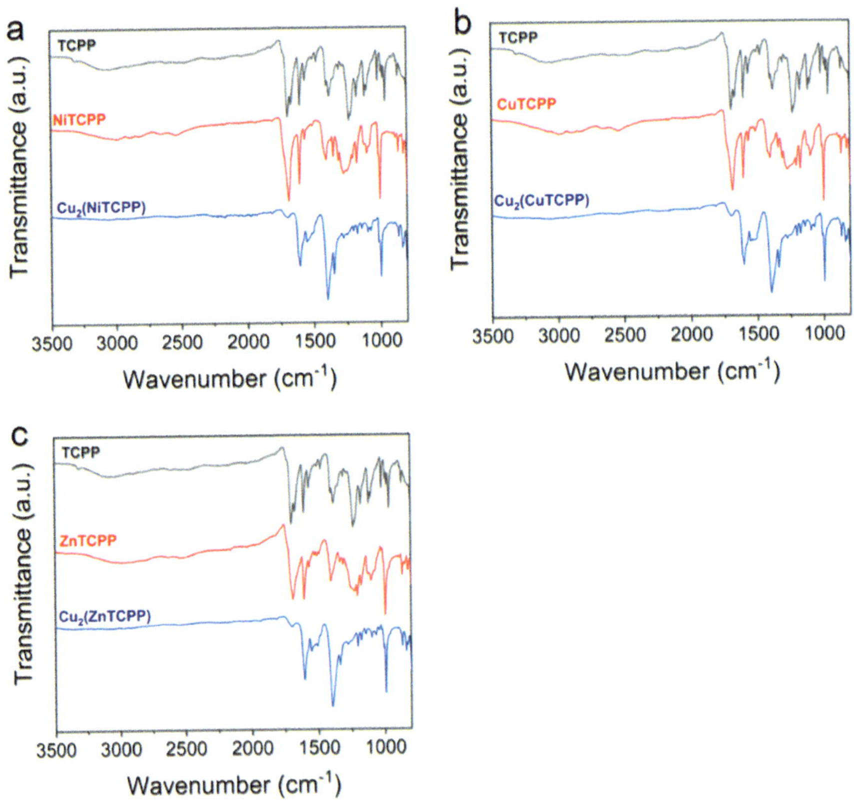

Fig. 4.6 Infrared spectra of H_4(NiTCPP) (**a**), Cu_2(CuTCPP) (**b**), and Cu_2(ZnTCPP) (**c**)

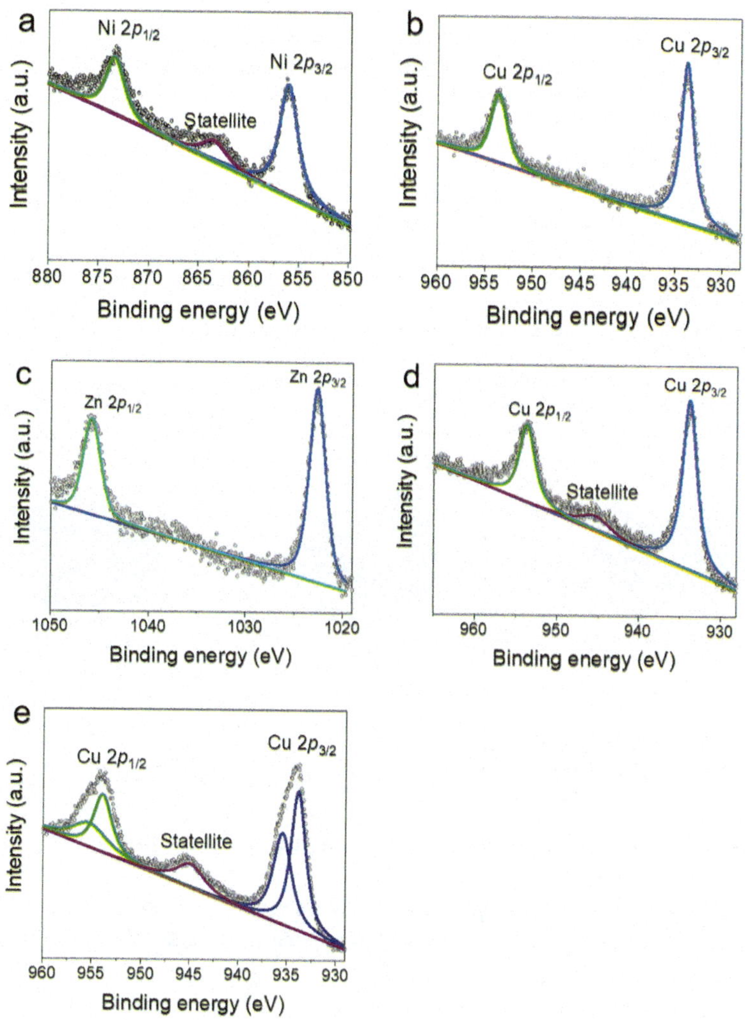

Fig. 4.7 X-ray photoelectron spectra of (**a**) Ni 2*p* and (**b**) Cu 2*p* regions in Cu_2(NiTCPP) nanosheets, (**c**) Zn 2*p* and (**d**) Cu 2*p* regions in Cu_2(ZnTCPP) nanosheets, (**e**) Cu 2*p* regions in Cu_2(CuTCPP) nanosheets

because of its 2D layered structure. The limited bending radius (R) of the 8-μm-thick membrane is less than 1 mm (Fig. 4.12), which is significantly smaller than that of a 500-nm-thick graphene membrane obtained by chemical vapor deposition ($R \approx 2.3$ mm) [20]. Furthermore, we confirmed that the membrane returns to its original flat state after repeated bending and relaxation operations of the membrane at least 20 times. We performed a three-point bending test of the membranes. Notably, the flexural moduli (approximately 2.8 GPa) were comparable with that of typical commercial polymers, such as polycarbonate (PC), poly(butylene terephthalate)

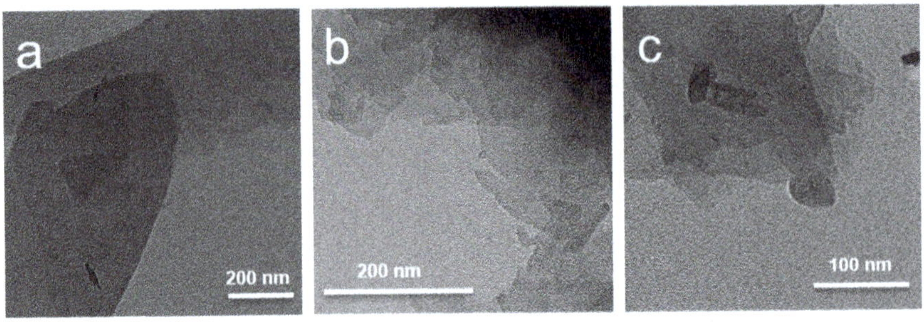

Fig. 4.8 Transmission electron microscope images of Cu_2(NiTCPP) (**a**), Cu_2(CuTCPP) (**b**), and Cu_2(ZnTCPP) (**c**) nanosheets

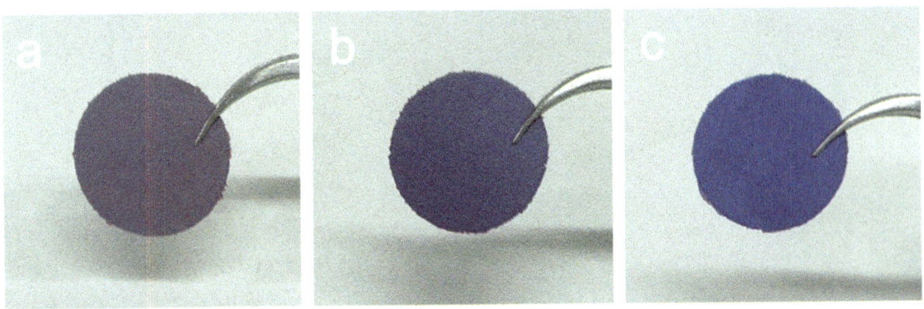

Fig. 4.9 Photographs of the Cu_2(NiTCPP) (**a**), Cu_2(CuTCPP) (**b**), and Cu_2(ZnTCPP) (**c**) Cu_2(NiTCPP) free-standing membrane

(PBT), poly(ethylene terephthalate) (PET), and poly(ether ether ketone) (PEEK) [21].

Room-temperature in-plane proton conductivity (σ) of a Cu_2(MTCPP) membranes with a thickness of 60 μm was measured using the a.c. impedance method. As shown in Fig. 4.13, the size of the high-frequency semicircles in the Nyquist plot decreased as the relative humidity (RH) increased from 75% RH to 95% RH. This tendency indicates that the protons originating from the uncoordinated carboxylic groups [22] act as main charge carriers through Grotthuss hopping between the water molecules in the interlayer spacing. Figure 4.13b shows the Nyquist dot plots, fitting equivalent circuits and fitting curves of different films at 95% humidity. Through calculation, the conductivity of the three films at 25 degrees Celsius and 95% humidity is 8.85×10^{-6}, 5.64×10^{-6}, and 1.50×10^{-5} S cm^{-1} of Cu_2(NiTCPP), Cu_2(CuTCPP), and Cu_2(ZnTCPP), respectively. Figure 4.14 shows the humidity-dependent curve of different membranes. It can be seen from the figure that proton conductivity meets Cu_2(NiTCPP) < Cu_2(CuTCPP) < Cu_2(ZnTCPP) relationship under the same humidity. There are two reasons for the difference in proton conduction. The first is the difference in the hydration coordination ability of porphyrin central metals [23]. The coordination ability of Zn and Ni with water is greater than that of Cu, which means that under the same humidity, Zn and Ni can

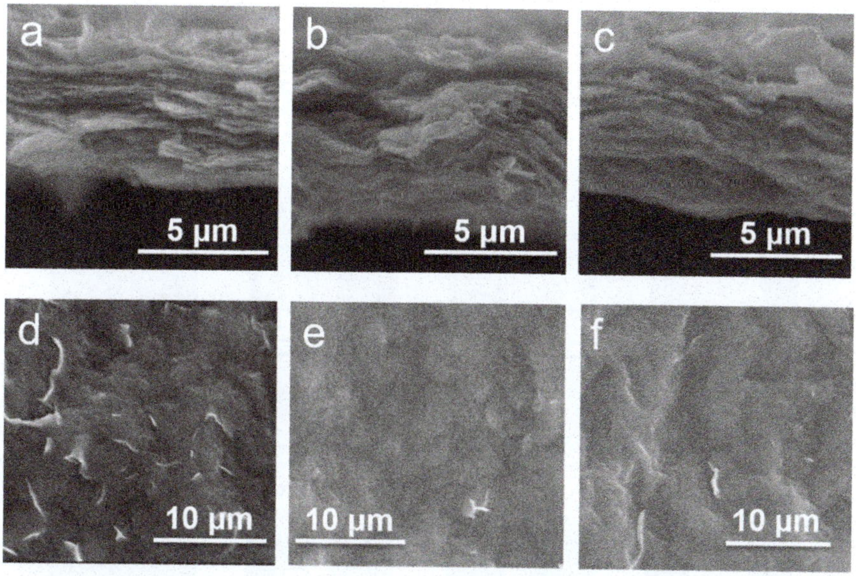

Fig. 4.10 Cross-sectional and surface scanning electron microscope images of the Cu_2(NiTCPP) (**a**, **d**), Cu_2(CuTCPP) (**b**, **e**), and Cu_2(ZnTCPP) (**c**, **f**) free-standing membrane

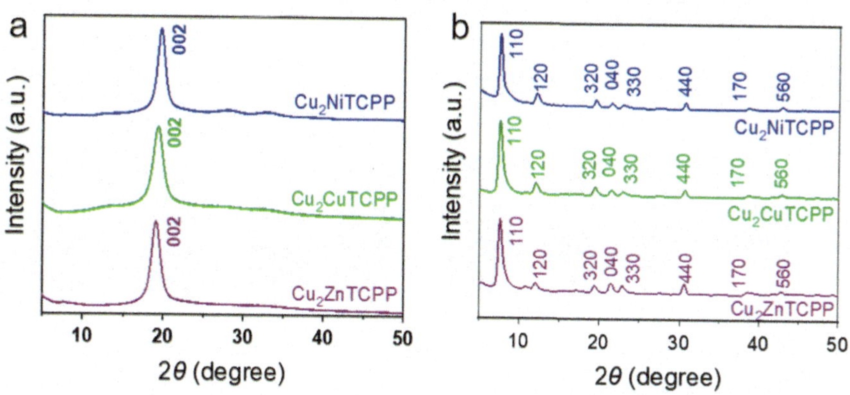

Fig. 4.11 Out-of-plane **a** and in-plane **b** X-ray diffraction (XRD) patterns of the Cu_2(NiTCPP) (blue), Cu_2(CuTCPP) (green), and Cu_2(ZnTCPP) (purple) free-standing membrane

adsorb more water molecules to form dense hydrogen bond network. The second reason is the difference in layer spacing. By comparing the XRD curves of the three envelopes, as shown in Fig. 4.15, it can be found that the layer spacing in the three membranes meet (Cu_2(NiTCPP) < Cu_2(CuTCPP) < Cu_2(ZnTCPP)). Larger layer spacing not only provides larger proton conduction channels, which is conducive for water molecules to carry protons quickly, but also accommodates more water molecules between layers to form a denser hydrogen bond network, which provides

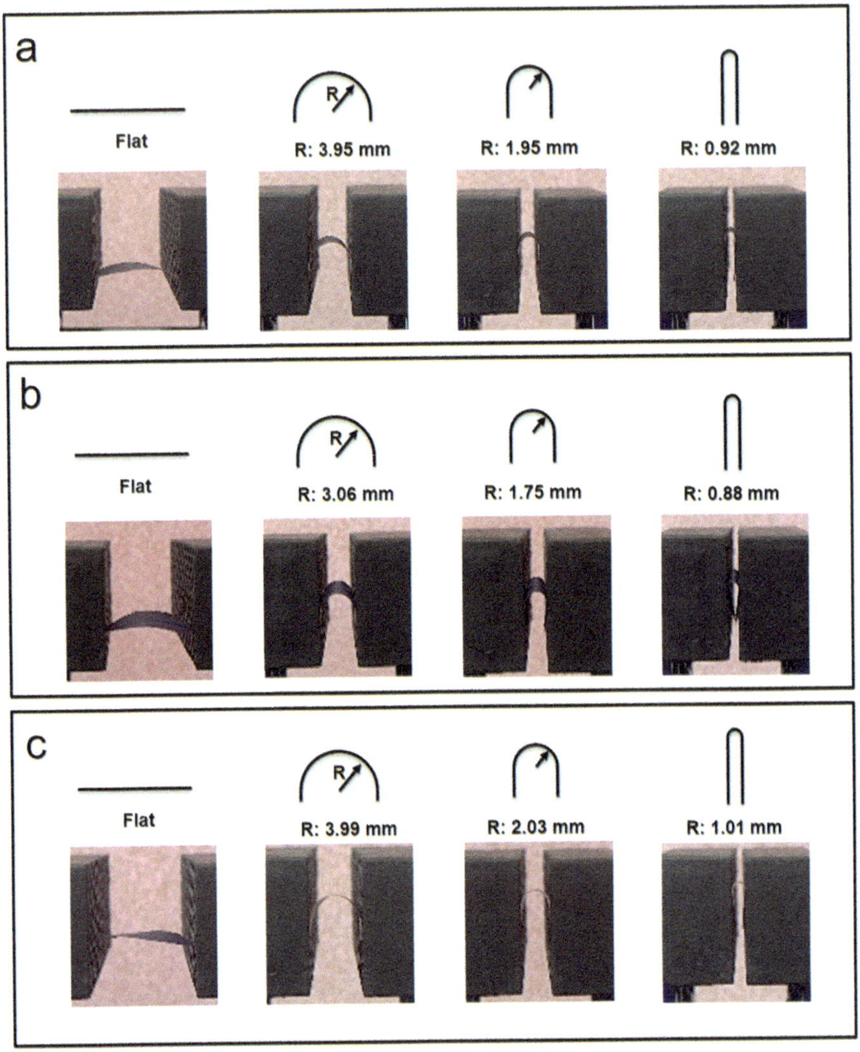

Fig. 4.12 Photographs of the 8-μm-thick membrane **a** Cu_2(NiTCPP), **b** Cu_2(CuTCPP), and **c** Cu_2(ZnTCPP) during the bending operation

more paths for proton conduction. Figure 4.16a displays the temperature dependence of the Nyquist plots of the Cu_2(NiTCPP), (b) Cu_2(CuTCPP), and (c) Cu_2(ZnTCPP) membrane, respectively. For all membranes, the σ value steadily increased with increasing temperature and eventually reached 4.53 × 10^{-5}, 2.00 × 10^{-5}, and 4.73 × 10^{-5} S cm^{-1} at 45 °C in Cu_2(NiTCPP), (b) Cu_2(CuTCPP), and Cu_2(ZnTCPP) membranes, respectively. Figure 4.16b shows the Arrhenius plots for the σT values of the three membranes. Each activation energy (E_a) was estimated to be 0.51, 0.42, and 0.37 eV in Cu_2(NiTCPP), (b) Cu_2(CuTCPP), and (c) Cu_2(ZnTCPP) membranes,

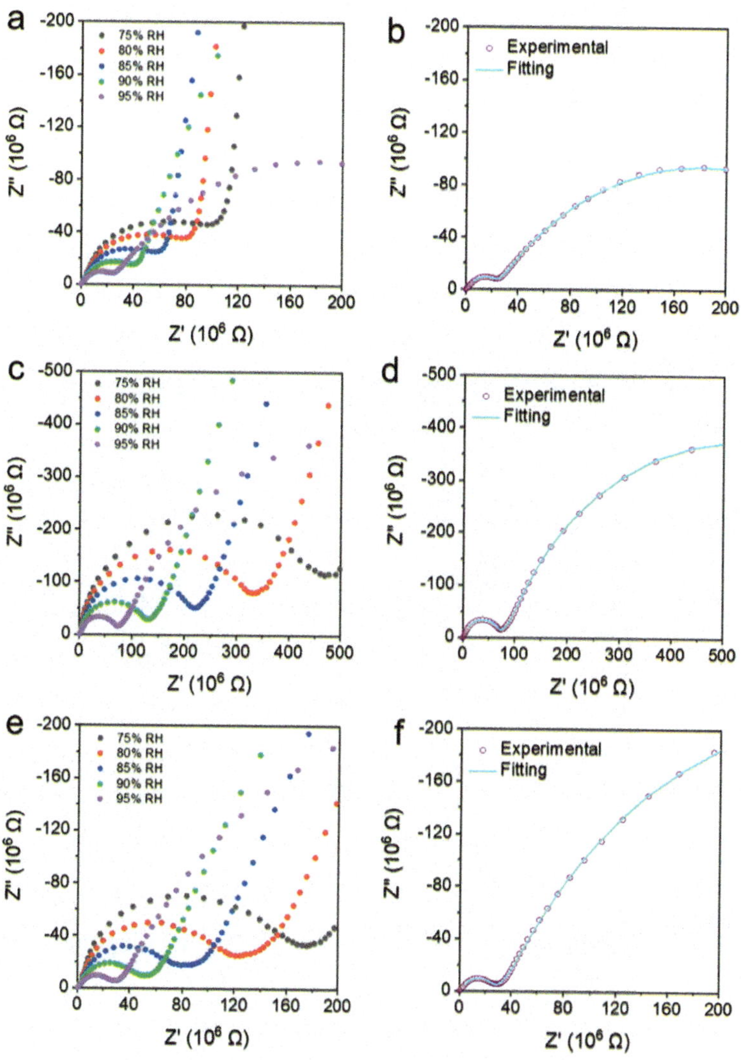

Fig. 4.13 Relative humidity-dependent Nyquist plots and the fitting results at 25 °C and 95% RH of the Cu_2(NiTCPP) (**a**, **b**), Cu_2(CuTCPP) (**c**, **d**), and Cu_2(ZnTCPP) (**e**, **f**) membranes

respectively, using the Arrhenius equation given by $\sigma T = A\ \exp(-\ E_a/k_B T)$ (A: pre-exponential factor; k_B: Boltzmann constant). The E_a values suggest that proton-conducting mechanism Grotthuss and Vehicle mechanism almost play the same effect [24].

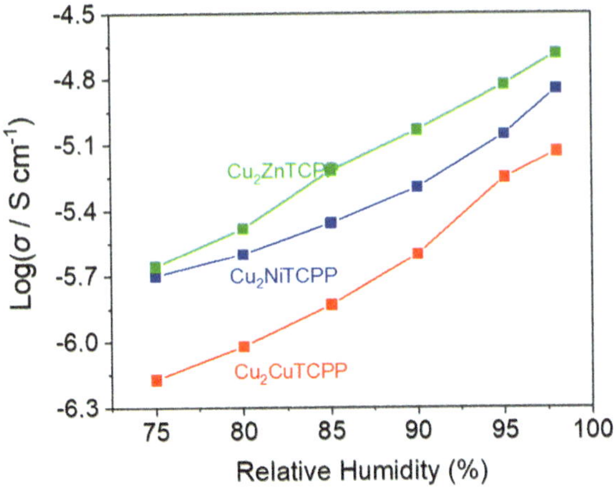

Fig. 4.14 Proton conductivity (σ) of the Cu_2(NiTCPP) (blue line), Cu_2(CuTCPP) (red line), and Cu_2(ZnTCPP) (green line) membranes at 25 °C under various relative humidities

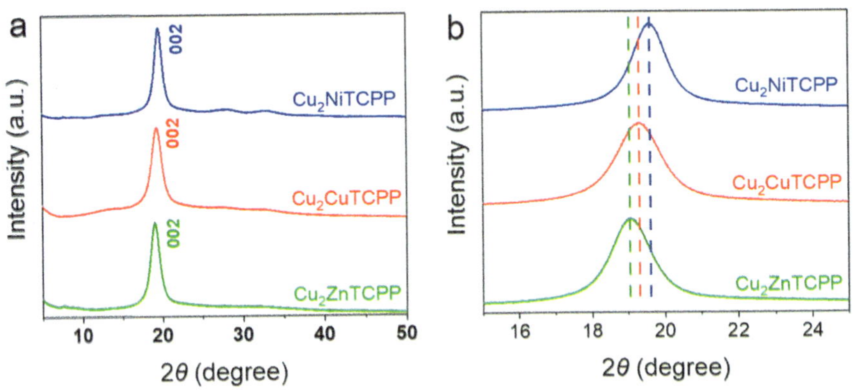

Fig. 4.15 Out-of-plane X-ray diffraction (XRD) patterns of the Cu_2(NiTCPP) (blue line), Cu_2(CuTCPP) (red line), and Cu_2(ZnTCPP) (green line) membranes

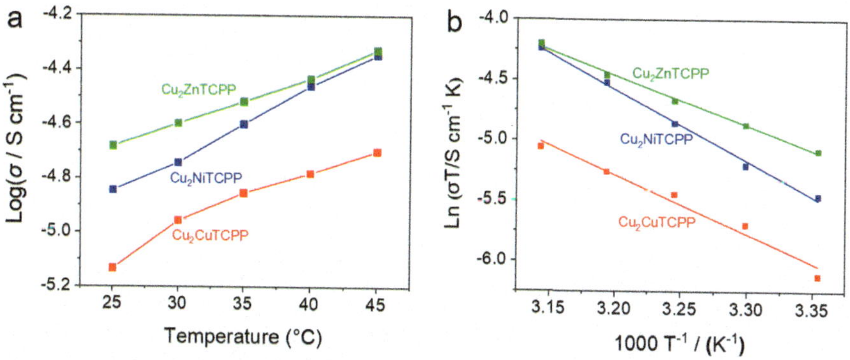

Fig. 4.16 Temperature dependence of proton conductivity **a** and Arrhenius plots of σ **b** of the Cu_2(NiTCPP) (blue line), Cu_2(CuTCPP) (red line), and Cu_2(ZnTCPP) (green line) membranes

4.4 Conclusion

In conclusion, we fabricated highly flexible free-standing membranes composed of 2D CPs, Cu_2(NiTCPP), Cu_2(CuTCPP), and Cu_2(ZnTCPP). The membrane was highly oriented and had excellent mechanical properties owing to the weak interlayer interactions of van der Waals type. The σ value of the membrane reached 4.53×10^{-5}, 2.00×10^{-5}, and 4.73×10^{-5} S cm^{-1} at 45 °C and 98% RH condition. This is the first study to fabricate a series of flexible CP free-standing membranes. The prominent advantage of the present membrane for future flexible protonic devices is that it can be fabricated free from any substrate and shows excellent bending stability even without blending with the polymer matrix. Thus, this study will offer the opportunity to extend the range of feasible applications of CPs in protonic devices.

References

1. (a) V. Rubio-Giménez, S. Tatay, C. Martí-Gastaldo, Chem. Soc. Rev. **49,** 5601–5638 (2020); (b) J. Liu, M. Zheng, S. Wu, L. Zhang, Coord. Chem. Rev. **483,** 215084 (2023); (c) Q. Dang, W. Zhang, X. Li, M. Hu, Mol. Syst. Des. Eng. **4,** 531–544 (2019); (d) H. Maeda, R. Sakamoto, H. Nishihara, Coord. Chem. Rev. **346,** 139–149 (2017)
2. (a) M. Hao, M. Qiu, H. Yang, B. Hu, X. Wang, Sci. Total Environ.**760,** 143333 (2021); (b) S. Mallakpour, E. Nikkhoo, C.M. Hussain, Coord. Chem. Rev. **451,** 214262 (2022); (c) W. Fan, X. Zhang, Z. Kang, X. Liu, D. Sun, Coord. Chem. Rev. **443,** 213968 (2021); (d) J. Canivet, A. Fateeva, Y. Guo, B. Coasne, D. Farrusseng, Chem. Soc. Rev. **43,** 5594–5617 (2014)
3. (a) Z. Zheng, H.L. Nguyen, N. Hanikel, K.K.Y. Li, Z. Zhou, T. Ma, O.M. Yaghi, Nat. Protoc.**18,** 136–156 (2023); (b) N. Hanikel, M.S. Prévot, O.M. Yaghi, Nat. Nanotechnol. **15,** 348–355 (2020)
4. (a) T. Yamada, K. Otsubo, R. Makiura, H. Kitagawa, Chem. Soc. Rev. **42,** 6655–6669 (2013); (b) Y. Ye, L. Gong, S. Xiang, Z. Zhang, B. Chen, Adv. Mater. **32,** 1907090 (2020); (c) L. Feng, H.B. Hou, H. Zhou, Dalton Trans. **49,** 17130–17139 (2020)

5. L.C. An, X. Li, Z.G. Li, Q. Li, P.J. Beldon, F.F. Gao, Z.Y. Li, S. Zhu, L. Di, S. Zhao, J. Zhu, D. Comboni, I. Kupenko, W. Li, U. Ramamurty, X.H. Bu, Nat. Commun. **13**, 6645 (2022)
6. (a) M. Pandey, N. Kumari, S. Nagamatsu, S.S. Pandey, J. Mater. Chem. C **7,** 13323–13351 (2019); (b) D. Li, W.Y. Lai, Y. Z. Zhang, W. Huang, Adv. Mater. **30,** 1704738 (2018); (c) X. Fan, W. Nie, H. Tsai, N. Wang, H. Huang, Y. Cheng, R. Wen, L. Ma, F. Yan, Y. Xia, Adv. Sci. **6,** 1900813 (2019)
7. (a) O. Shekhah, J. Liu, R.A. Fischer, C. Wöll, Chem. Soc. Rev. **40,** 1081–1106 (2011); (b) Y. Liu, Y. Ban, W. Yang, Adv. Mater. **29,** 1606949 (2017); (c) Z. Kang, L. Fan, D. Sun, J. Mater. Chem. A **5,** 10073–10091 (2017)
8. (a) E. Adatoz, A.K. Avci, S. Keskin, Sep. Purif. Technol. **152,** 207–237 (2015); (b) Y. Shi, B. Liang, R.B. Lin, C. Zhang, B. Chen, Trends Chem. **2,** 254–269 (2020); (c) J. Dechnik, J. Gascon, C. J. Doonan, C. Janiak, C.J. Sumby, Angew. Chem. Int. Ed. Engl. **56,** 9292–9310 (2017); (d) M. Kalaj, K. C. Bentz, S. Ayala Jr, J.M. Palomba, K.S. Barcus, Y. Katayama, S.M. Cohen, Chem. Rev. **120,** 8267–8302 (2020); (e) X. Li, Y. Liu, J. Wang, J. Gascon, J. Li, B. Van der Bruggen, Chem. Soc. Rev. **46,** 7124–7144 (2017)
9. S. Jia, D. Ji, L. Wang, X. Qin, S. Ramakrishna, Small Struct. **3**, 2100222 (2022)
10. H.S. Sasmal, H.B. Aiyappa, S.N. Bhange, S. Karak, A. Halder, S. Kurungot, R. Banerjee, Angew. Chem. **130**, 11060–11064 (2018)
11. (a) M. Đaković, M. Borovina, M. Pisačić, C. B. Aakeröy, Ž. Soldin, B.M. Kukovec, I. Kodrin, Angew. Chem. Int. Ed. **57,** 14801–14805 (2018), Angew. Chem. **130,** 15017–15021 (2018); (b) M. Pisacic, I. Biljan, I. Kodrin, N. Popov, Z. Soldin, M. Đaković, Chem. Mater. 33, 3660–3668 (2021); (c) B. Bhattacharya, A.A. Michalchuk, D. Silbernagl, M. Rautenberg, T. Schmid, T. Feiler, K. Reimann, A. Ghalgaoui, H. Sturm, B. Paulus, F. Emmerling, Angew. Chem. Int. Ed. **59,** 5557–5561 (2020), Angew. Chem. **59,** 5557–5561 (2020)
12. M. Tian, F. Pei, M. Yao, Z. Fu, L. Lin, G. Wu, G. Xu, H. Kitagawa, X. Fang, Energy Storage Mater. **21**, 14–21 (2019)
13. J. Lu, Y. Yoshida, M. Maesato, H. Kitagawa, Angew. Chem. Int. Ed. **61,** e202213077 (2022), Angew. Chem. **134,** e202213077 (2022)
14. (a) G. Granados-Oliveros, E.A. Páez-Mozo, F.M. Ortega, C. Ferronato, J.M. Chovelon, Appl. Catal. B **89,** 448–454 (2009); (b) P.K. Sonkar, K. Prakash, M. Yadav, V. Ganesan, M. Sankar, R. Gupta, D.K. Yadav, J. Mater. Chem. **5,** 6263–6276 (2017); (c) K. Jeong, J.M. Kim, S.H. Kim, G.Y. Jung, J. Yoo, S.H. Kim, S.K. Kwak, S.Y. Lee, Adv. Funct. Mater. **29,** 1806937 (2019)
15. (a) G. Xu, T. Yamada, K. Otsubo, S. Sakaida, H. Kitagawa, J. Am. Chem. Soc. **134,** 16524–16527 (2012); (b) G. Xu, K. Otsubo, T. Yamada, S. Sakaida, H. Kitagawa, J. Am. Chem. Soc. **135,** 7438–7441 (2013)
16. L. Wang, P. Jin, J. Huang, H. She, Q. Wang, A.C.S. Sustain, Chem. Eng. **7**, 15660–15670 (2019)
17. C. Berrios, G.I. Cardenas-Jiron, J.F. Marco, C. Gutierrez, M.S. Ureta-Zanartu, J. Phys. Chem. A **111**, 2706–2714 (2007)
18. (a) H. Ang, L. Hong, ACS Appl. Mater. Inter. **9,** 28079–28088 (2017). (b) M. Cong, X. Chen, K. Xia, X. Ding, L. Zhang, Y. Jin, Y. Gao, L. Zhang, J. Mater. Chem. A **9,** 4673–4678 (2021). (c) L.Y. Huang, X. Hu, H.Y. Shan, L. Yu, Y.X. Gu, A.J. Wang, D. Shan, P.X. Yuan, J. J. Feng, Sensor. Actuat. B Chem. **344,** 130300 (2021)
19. Y. Zhou, S. Chen, S. Xi, Z. Wang, P. Deng, F. Yang, Y. Han, Y. Pang, B.Y. Xia, Cell Rep. Phys. Sci. **1**, 100182 (2020)
20. J. Yun, D. Kim, G. Lee, J.S. Ha, Carbon **79**, 156–164 (2014)
21. K. Van de Velde, P. Kiekens, Polym. Test. **20**, 885–893 (2001)
22. Y.N. Zhou, L.L. Liu, Q.W. Liu, X.X. Liu, M.Z. Feng, L. Wang, Z.G. Sun, Y.Y. Zhu, X. Zhang, C.Q. Jiao, Inorg. Chem. **60**, 17303–17314 (2021)
23. (a) K.I. Yamashita, K. Furutani, T. Ogawa, Asian J. Org. Chem. **10,** 1192–1197 (2021). (b) P. Hambright, ChemComm. (London) **9,** 470–471 (1967)
24. D. Marx, ChemPhysChem **7**, 1848–1870 (2006)

Chapter 5
General Conclusion

Abstract This Ph.D. thesis presents the fabrication of flexible, free-standing 2D coordination polymer membranes with specific proton-conducting properties. The study explores the principles of proton conduction, fabricates CP membranes, and demonstrates robust proton conduction under mechanical stress. A record rectification ratio was achieved with a heterogeneous membrane, and the membranes show excellent bending stability, indicating potential for flexible protonic devices without a polymer matrix.

Keywords 2D coordination polymers · Proton conduction · Free-standing membranes · Mechanical stress · Flexible protonic devices

This Ph.D. thesis fabricated a series of highly flexible free-standing membranes composed of 2D coordination polymers (CPs). Specific proton-conducting behavior was performed in these membranes.

In Chap. 1, the author mainly focused on the research background. Firstly, the author describes proton conduction's basic principles, mechanisms, and advantages. Then, the author describes the way to fabricate CPs' membranes. Finally, the author explained the specific proton-conducting behavior and its mechanisms and applications.

Chapter 2, "High-Performance All-Solid-State Proton Rectifier Using a Heterogeneous Membrane Composed of Coordination Polymer and Layered Double Hydroxide," is focused on the fabrication of all-solid-state proton rectifier by the facile mechanical lamination of two free-standing membranes, namely, a proton-conducting PCP, Cu_2 (CuTCPP), and hydroxide ion-conducting LDH, Mg-Al-LDH(NO_3). The I–V characteristics provide a rectification ratio as high as > 200 at 25 °C under 90% RH, the highest recorded value.

In Chap. 3, "Robust Proton Conduction against Mechanical Stress in Flexible Free-Standing Membrane Composed of Two-Dimensional Coordination Polymer," we fabricated highly flexible free-standing membranes composed of 2D CP, Cu_2 (NiTCPP). The membrane is highly oriented and has excellent mechanical properties by virtue of weak interlayer interactions of the van der Waals type. The proton

J. Lu, *Novel Proton Conducting in Free-Standing Coordination Polymer Membranes*, Springer Theses, https://doi.org/10.1007/978-981-97-7913-0_5

conductivity of the membrane remains almost unchanged even by applying bending stress. The proton-conducting properties and structural features in different bending states strongly indicate that the proton-conducting pathway through the hydrogen-bonding network is kept intact during the bending operation. This is the first study to demonstrate the robust proton conduction of 2D CP against the applied bending stress.

Following Chap. 4, "Proton Conducting Behavior in Flexible Coordination Polymer Free-Standing Membranes," we fabricated highly flexible free-standing membranes composed of 2D CPs, Cu_2 (NiTCPP), Cu_2 (CuTCPP), and Cu_2 (ZnTCPP). The membrane was highly oriented and had excellent mechanical properties owing to the weak interlayer interactions of the van der Waals type. The σ value of the membrane reached 4.53×10^{-5}, 2.00×10^{-5}, and 4.73×10^{-5} S cm^{-1} at 45 °C and 98% RH condition. This is the first study to fabricate a series of flexible CP free-standing membranes. The prominent advantage of the present membrane for future flexible protonic devices is that it can be fabricated free from any substrate and shows excellent bending stability even without blending with the polymer matrix.

This is the first study to demonstrate the robust proton conduction of 2D CP against the applied bending stress. The prominent advantage of the present membrane for future flexible protonic devices is that it can be fabricated free from any substrate and shows excellent bending stability even without blending with a polymer matrix. Our present study will offer an opportunity to extend the range of feasible applications of CPs in protonic devices.

Curriculum Vitae

Jiangfeng Lu

Fujian Institute of Material Structure,

Chinese Academy of Sciences.

No. 8 Gaoxin Avenue, Shangjie Town,

Minhou County, Fuzhou, Fujian, China.

+ 86–137–8730–3030

Lujiangfeng812@gmail.com

Education

October 2019–September 2023:

Ph.D. in Chemistry, Kyoto University.

Supervisor: Professor Hiroshi Kitagawa.

J. Lu, *Novel Proton Conducting in Free-Standing Coordination Polymer Membranes*,
Springer Theses, https://doi.org/10.1007/978-981-97-7913-0

September 2017–July 2020:

M.Sc. in Materials Engineering, Central South University.

Supervisor: Dr Qiaozhen Sun.

September 2011–June 2017:

B.Sc. in Materials Science and Engineering, Shenyang University of Chemical Technology.

Employment

January 2024–present, Postdoctoral program at Fujian Institute of Material Structure, Chinese Academy of Sciences.

April 2021–September 2023, Fellowship for Future Human Resources Development to Create Materials Innovation at Graduate School of Science, Kyoto University.

Research Interest

Proton/electron-conducting metal-organic frameworks.

Coordination polymers' membrane.

Gas sensing.

SPRINGER NATURE

GPSR Compliance

The European Union's (EU) General Product Safety Regulation (GPSR) is a set of rules that requires consumer products to be safe and our obligations to ensure this.

If you have any concerns about our products, you can contact us on ProductSafety@springernature.com

In case Publisher is established outside the EU, the EU authorized representative is:

Springer Nature Customer Service Center GmbH
Europaplatz 3
69115 Heidelberg, Germany

The manufacturer's authorised representative in the EU is Springer Nature Customer Service Centre GmbH, Europaplatz 3, 69115 Heidelberg, Germany. If you have any concerns regarding our products, please contact ProductSafety@springernature.com

Printed and bound by CPI Group (UK) Ltd, Croydon, CR0 4YY
30/09/2026
02225308-0002